About Access Archaeology

Access Archaeology offers a different publishing model for specialist academic material that might traditionally prove commercially unviable, perhaps due to its sheer extent or volume of colour content, or simply due to its relatively niche field of interest. This could apply, for example, to a PhD dissertation or a catalogue of archaeological data.

All *Access Archaeology* publications are available in open-access e-pdf format and in print format. The open-access model supports dissemination in areas of the world where budgets are more severely limited, and also allows individual academics from all over the world the opportunity to access the material privately, rather than relying solely on their university or public library. Print copies, nevertheless, remain available to individuals and institutions who need or prefer them.

The material is refereed and/or peer reviewed. Copy-editing takes place prior to submission of the work for publication and is the responsibility of the author. Academics who are able to supply print-ready material are not charged any fee to publish (including making the material available in open-access). In some instances the material is type-set in-house and in these cases a small charge is passed on for layout work.

Our principal effort goes into promoting the material, both in open-access and print, where *Access Archaeology* books get the same level of attention as all of our publications which are marketed through e-alerts, print catalogues, displays at academic conferences, and are supported by professional distribution worldwide.

Open-access allows for greater dissemination of academic work than traditional print models could ever hope to support. It is common for an open-access e-pdf to be downloaded hundreds or sometimes thousands of times when it first appears on our website. Print sales of such specialist material would take years to match this figure, if indeed they ever would.

This model may well evolve over time, but its ambition will always remain to publish archaeological material that would prove commercially unviable in traditional publishing models, without passing the expense on to the academic (author or reader).

Anglo-Saxon Crops and Weeds

A Case Study in Quantitative Archaeobotany

Mark McKerracher

ARCHAEOPRESS PUBLISHING LTD
Summertown Pavilion
18-24 Middle Way
Summertown
Oxford OX2 7LG

www.archaeopress.com

ISBN 978-1-78969-192-4
ISBN 978-1-78969-193-1 (e-Pdf)

© M McKerracher and Archaeopress 2019

Cover: Cowslips on Magdalen Hill Down, Hampshire © Mark McKerracher

Printed and bound in Great Britain by
Marston Book Services Ltd, Oxfordshire

All rights reserved. No part of this book may be reproduced, stored in retrieval system, or transmitted, in any form or by any means, electronic, mechanical, photocopying or otherwise, without the prior written permission of the copyright owners.

This book is available direct from Archaeopress or from our website www.archaeopress.com

To Mr Snow
who taught me history and maths
and told me to roar.

Two out of three isn't bad.

Contents

List of figures ii

List of tables v

Abbreviations and notes vii

Acknowledgements viii

Chapter 1: Seeds of Change 1

Chapter 2: Describing the Data 11

Chapter 3: Surveying the Species 28

Chapter 4: Defining the Deposits 35

Chapter 5: Counting the Crops 64

Chapter 6: The Witness of Weeds 96

Chapter 7: More than the Sum of their Parts 128

Appendix 1: Key Parameters 131

Appendix 2: Key Metadata 133

Appendix 3: Gazetteer of Sites 146

Appendix 4: Inventory of Samples 152

Appendix 5: Inventory of Plant Taxa 179

Bibliography 187

List of figures

All maps contain OS data © Crown copyright and database right 2017.

Figure 1 - Modern administrative geography of the case study regions.................................12

Figure 2 - National Character Areas represented within the case study regions. 16

Figure 3 - Proportions of analyses undertaken by different archaeobotanists within the project dataset. 19

Figure 4 - Geographical distribution of sites represented in the project dataset. 25

Figure 5 – Tripolar graph of grain : rachis : weed seed ratios in ethnographic control samples analysed by Jones 1990: 93-96. 38

Figure 6 – Idealized interpretation of grain : rachis : weed seed ratios in terms of crop processing products and by-products, after Jones 1990. 39

Figure 7 – Tripolar graph of grain : rachis : weed seed ratios of free-threshing cereal samples in the project dataset. 39

Figure 8 – Model scattergraph for the visual interpretation of discriminant analysis results, after Jones 1987: 315, Figure 1. 44

Figure 9 – Discriminant analysis scattergraph of free-threshing cereal samples. 47

Figure 10 – Discriminant analysis scattergraph of free-threshing cereal samples deemed to have a classification compatible with their basic composition ratios. The symbols used here denote the classifications obtained through the ratio analysis above (Table 6)...52

Figure 11 – Average density of charred plant remains in grain-rich free-threshing cereal samples, grouped chronologically. 54

Figure 12 – Average density of charred plant remains in grain-rich free-threshing cereal samples, grouped according to parent feature type. 55

Figure 13 – Discriminant analysis scattergraph of glume wheat samples. 59

Figure 14 – Discriminant analysis scattergraph of mixed cereal type samples. 62

Figure 15 – Phased presence analysis of barley and wheat remains, by assemblages. 68

Figure 16 – Phased presence analysis of free-threshing and glume wheat remains, by assemblages. ...68

Figure 17 – Phased presence analysis of different glume wheat remains, by assemblages. 69

Figure 18 – Phased presence analysis of oat and rye remains, by assemblages. 69

Figure 19 – Phased presence analysis of barley and wheat remains, by samples. 71

Figure 20 – Phased presence analysis of free-threshing and glume wheat remains, by samples. 71

Figure 21 – Phased presence analysis of different glume wheat remains, by samples. 72

Figure 22 – Phased presence analysis of oat and rye remains, by samples. 72

Figure 23 – Regional presence analysis of barley remains, by samples. 75

Figure 24 – Regional presence analysis of wheat remains, by samples. 75

Figure 25 – Regional presence analysis of free-threshing wheat remains, by samples. 76

Figure 26 – Regional presence analysis of spelt remains, by samples. 76

Figure 27 – Regional presence analysis of emmer remains, by samples. 77

Figure 28 – Regional presence analysis of oat remains, by samples. 77

Figure 29 – Regional presence analysis of rye remains, by samples. 78

Figure 30 – Percentage of rye grains in free-threshing product samples, grouped chronologically. 84

Figure 31 - Percentage of free-threshing wheat grains in free-threshing product samples, grouped chronologically. 85

Figure 32 - Percentage of barley grains in free-threshing product samples, grouped chronologically. 86

Figure 33 - Percentage of oat grains in free-threshing product samples, grouped chronologically. 87

Figure 34 – Interpolated map of percentage of barley grains in free-threshing product samples. 87

Figure 35 - Interpolated map of percentage of free-threshing wheat grains in free-threshing product samples. 88

Figure 36 - Interpolated map of percentage of oat grains in free-threshing product samples. 88

Figure 37 - Interpolated map of percentage of rye grains in free-threshing product samples. 89

Figure 38 – Distribution of samples in correspondence analysis of free-threshing product samples and constituent weed species. 105

Figure 39 - Distribution of species in correspondence analysis of free-threshing product samples and constituent weed species. 106

Figure 40 - Distribution of samples in correspondence analysis of free-threshing product samples and constituent weed species (excluding <WST2>). 107

Figure 41 - Distribution of species in correspondence analysis of free-threshing product samples and constituent weed species (excluding <WST2>). 108

Figure 42 - Distribution of samples in correspondence analysis of free-threshing product samples and constituent weed species, coded by crop processing classification. 110

Figure 43 - Distribution of samples in correspondence analysis of free-threshing product samples and constituent weed species, coded by flowering habits of weeds. 111

Figure 44 - Distribution of samples in correspondence analysis of free-threshing product samples and constituent weed species, coded by sowing time classification of samples. 112

Figure 45 - Distribution of samples in correspondence analysis of free-threshing product samples and constituent weed species, coded by life history of weeds. 113

Figure 46 - Distribution of samples in correspondence analysis of free-threshing product samples and constituent weed species, coded by moisture preferences of weeds. 114

Figure 47 - Distribution of samples in correspondence analysis of free-threshing product samples and constituent weed species, coded by nitrogen preferences of weeds. 115

Figure 48 - Distribution of samples in correspondence analysis of free-threshing product samples and constituent weed species, coded by acidity preferences of weeds. 116

Figure 49 - Distribution of samples in correspondence analysis of free-threshing product samples and constituent weed species, divided into broad compositional groups. 118

Figure 50 - Distribution of samples in correspondence analysis of free-threshing product samples and constituent weed species, coded by percentage of rye grains amongst grain content. 121

Figure 51 - Distribution of samples in correspondence analysis of free-threshing product samples and constituent weed species, coded by percentage of oat grains amongst grain content. 122

Figure 52 – Distribution of samples in correspondence analysis of free-threshing product samples and constituent weed species, coded by percentage of barley grains amongst grain content. 123

Figure 53 - Distribution of samples in correspondence analysis of free-threshing product samples and constituent weed species, coded by percentage of free-threshing wheat grains amongst grain content. ...124

List of tables

Table 1 – Representation of archaeobotanists in the project dataset...20

Table 2 - Chronological distribution of assemblages and samples in the project dataset.........................22

Table 3 - Geographical distribution of assemblages and samples in the project dataset, in terms of National Character Areas. 24

Table 4 – Traditional and new scientific nomenclature of cereals (after Cappers and Neef: 15-16); the traditional names are used in this book. 27

Table 5 – Characterisation of crop processing products and by-products according to constituent proportions of grain, chaff and weed seed (after Jones 1990: 93-96). 38

Table 6 - Crop processing analysis of free-threshing cereal samples by basic composition ratios. 40

Table 7 – Crop processing analysis of free-threshing cereal samples by discriminant analysis of weed seed types (discriminant functions rounded to three decimal places). 45

Table 8 - Theoretical compatibility of ratio and discriminant analysis crop processing classifications.48

Table 9 - Combined interpretation of crop processing analyses. 49

Table 10 – Chronological distribution of all samples compared with free-threshing grain-rich product (USG/FSP) samples. 53

Table 11 – Phased distribution of samples by feature type. 54

Table 12 - Relative proportions of grain, glume base and weed seed in samples dominated by glume wheats. 58

Table 13 - Crop processing analysis of glume wheat samples by discriminant analysis of weed seed types (discriminant functions rounded to three decimal places). 58

Table 14 - Relative proportions of grain, chaff and weed seed in mixed cereal samples. 61

Table 15 - Crop processing analysis of mixed cereal samples by discriminant analysis of weed seed types (discriminant functions rounded to three decimal places). 61

Table 16 – Abundance and average density of charred plant remains in mixed cereal samples.............63

Table 17 – Phased presence analysis of cereal taxa by assemblages...66

Table 18 – Phased presence analysis of cereal taxa by samples..70

Table 19 – Presence analysis of cereal taxa by samples, in terms of National Character Areas (including only NCAs with more than ten samples). 74

Table 20 – Relative proportions of taxa amongst grain in free-threshing grain-rich product samples. 80

Table 21 – Chronological distribution of free-threshing grain-rich product samples..............................81

Table 22 – Geographical distribution of free-threshing grain-rich product samples. 82

Table 23 - Relative proportions of cereal taxa, by glume bases and grains, in glume wheat samples. ...90

Table 24 - Relative proportions of cereal taxa amongst glume bases in mixed cereal samples..............91

Table 25 - Relative proportions of cereal taxa amongst rachis segments in mixed cereal samples........92

Table 26 - Relative proportions of cereal taxa amongst grains in mixed cereal samples. 93

Table 27 - Relationships between the flowering onset and duration of annual arable weeds and the sowing times of crops (after Bogaard *et al.* 2001: 1175, Table 3). 97

Table 28 - Relative proportions of early/short-, late-, and long-flowering species in the free-threshing product samples. 99

Table 29 - Sowing seasonality analysis of free-threshing product samples, by discriminant analysis of weed flowering onset and duration (discriminant function rounded to three decimal places)............100

Table 30 - Species included in correspondence analysis, with selected metadata..................................103

Table 31 - Samples included in correspondence analysis, with selected metadata. 104

Table 32 - Assemblage-based presence analysis of weed species, showing those with at least 10% change between the Early and Mid Saxon periods. 126

Table 33 - Sample-based presence analysis of weed species, showing those with at least 10% change between the Early and Mid Saxon periods. 127

Abbreviations and notes

EHD	*English Historical Documents* (Whitelock 1979)
HE	Bede, *Historia Ecclesiastica Gentis Anglorum*
VCB	Bede, *Vita Sancti Cuthberti*
HER	Historic Environment Record
NCA	National Character Area
NGR	National Grid Reference
SFB	sunken-featured building
Beds	Bedfordshire
Berks	Berkshire
Bucks	Buckinghamshire
Cambs	Cambridgeshire
Glos	Gloucestershire
Oxon	Oxfordshire
Wilts	Wiltshire

All dates are AD unless stated otherwise.

Numbers are rounded to one decimal place unless stated otherwise.

The terms *Grubenhaus* (plural *Grubenhäuser*) and sunken-featured building are considered to be synonymous.

The maps in this book have been produced using two free resources: the QGIS package (http://www.qgis.org, accessed April 2017) and Ordnance Survey Open Data made available under the Open Government Licence (https://www.ordnancesurvey.co.uk/business-and-government/products/opendata-products.html, accessed April 2017).

Acknowledgements

The work upon which this book is based was originally undertaken as part of a doctoral research project at the University of Oxford, funded by the Arts and Humanities Research Council via the Block Grant Partnership scheme. This volume contains an edited and revised version of the quantitative archaeobotanical portions of my completed thesis (McKerracher 2014a). I would like to thank my supervisors Helena Hamerow and Amy Bogaard for their invaluable guidance and support during my doctoral studies, and my examiners Mark Robinson and Mark Gardiner for their insightful critique.

My gratitude also extends to the many analysts and curators who facilitated access to original specialist reports during the course of my research, including (with sincere apologies for inadvertent omissions): Trevor Ashwin, Polydora Baker, Debby Banham, Angela Batt, Ian Baxter, Paul Booth, Sarah Botfield, Stuart Boulter, Esther Cameron, Gill Campbell, Jo Caruth, Brian Clarke, Pam Crabtree, Sally Croft, Anne Davis, Denise Druce, Brian Durham, Val Fryer, Sally Gale, Dave Gilbert, Jenny Glazebrook, Jessica Grimm, Julie Hamilton, Sheila Hamilton-Dyer, Sarah Howard, Anne-Marie McCann, Maureen Mellor, Mick Monk, John Moore, Jacqui Mulville, Peter Murphy, Andrew Newton, Leonora O'Brien, Nigel Page, Ruth Pelling, Colin Pendleton, Steve Preston, Sarah Pritchard, Dale Serjeantson, Kirsty Stonell Walker, Karen Thomas, Fay Worley, and Julia Wise.

A final 'thank you' must go to Mum, Dad and Rachel – I couldn't have done any of this without all of you.

Chapter 1: Seeds of Change

Between the 7th and 9th centuries AD, much of Europe and the Mediterranean world experienced a step-change in agricultural productivity. Farmlands in the Germanic north-west witnessed the laying-out of new field systems, the introduction of intensive turf-manuring, and a growth in grain storage spaces, which implies the growth of surplus production (Hamerow 2002: 139–147). Cereal cultivation was expanding, with a particular emphasis on rye and oats (Behre 1992: 148–150; Hamerow 2002: 134–137; Henning 2014: 335–336). In contemporary Carolingian Francia, documentary evidence for a profusion of mills, breweries and bakeries – especially on monastic estates – similarly points to a significant upturn in the production of cereal goods (Lebecq 2000: 134). More generally, a range of geoarchaeological and palaeoenvironmental surveys demonstrate that scrub and woodland were cleared to open up new farmland in southern France, Byzantine Italy and the Iberian peninsula around the 8th century (Arthur *et al.* 2012: 445–449; Puy and Balbo 2013: 46–51; Ruas 2005: 401). Regarding pastoral farming, the evidence of excavated animal bones suggests a new emphasis on sheep husbandry in northern France around the same period (Crabtree 2010: 129). Within and around the Carolingian world, farming was in flux.

Similar, contemporary developments may likewise be traced in the archaeology of Anglo-Saxon England, where this period is known to archaeologists as the Mid Saxon period (c. 650–850). Excavated Mid Saxon rural settlements begin to show more regularity in plan than their Early Saxon precursors (c. 450–650). In particular, they increasingly incorporate ditched systems of paddocks and droveways for the closer management of livestock (Hamerow 2012: 72–73). Zooarchaeological evidence shows that the animals themselves were kept alive for longer, and therefore yielded more secondary products such as milk, plough-traction and wool (Crabtree 2010; Holmes 2014). Palaeoenvironmental evidence from numerous pollen cores, as well as sedimentary sequences, point to an expansion of arable land and cereal cultivation, once again between the 7th and 9th centuries (Rippon 2010). The growth of cereal surpluses in this period is further indicated by the reappearance in Britain, for the first time since the Roman period, of watermills, granaries and large stone-built grain ovens (Gardiner 2012; Hamerow 2012: 151–152; McKerracher 2014b).

In this way, it is increasingly being recognized that farming practices changed and became more productive in Mid Saxon England, as part of a wider process of agricultural development across the post-Roman world. To date, however, most studies of agrarian change in early medieval England have focused largely upon the broad outlines of change – such as the growth of crop husbandry at the expense of pastoralism – without considering the closer, more practical details of innovation, such as how cultivation practices developed to support the expansion of arable farming. A more fine-grained perspective can now be achieved, however, thanks to the massive growth of development-led archaeology in England over the past three decades and the concomitant increase in palaeoenvironmental sampling at these excavations (Bradley 2006).

The result of these twin trends is a rich but relatively little utilised national dataset of charred plant remains, chiefly characterised by cereal grains, cereal chaff and the seeds of arable weeds that were accidentally harvested with the crops (van der Veen *et al.* 2013). Over the same period, a combination of ethnographic, ecological and statistical research has been deployed by prehistorians to develop sensitive, quantitative archaeobotanical methods for the reconstruction of past crop husbandry regimes (Jones *et al.* 2010). These methods and models, being based upon functional ecological observations and the inherent characteristics of charred crop deposits, are as applicable to early medieval assemblages as they are to the Neolithic and Bronze Age material to which they have hitherto mostly been applied.

With that in mind, this monograph applies a range of quantitative methods to archaeobotanical data from Anglo-Saxon England, in order to shed more light on the agricultural innovations of the Mid Saxon period. This quantitative approach provides a complement to the more qualitative studies of the same topic which I have published elsewhere, and also fleshes out the statistical background to the dataset used in those publications (McKerracher 2016; 2018). In so doing, this book also proposes a standardised, repeatable set of protocols for the application of previously-developed, tried-and-tested quantitative methods, to facilitate their use in comparable future research projects. Procedures, variables, assumptions, decisions and data are therefore exposed as fully as possible in this book. Where appropriate, certain variables are also parameterised: defined as changeable values – parameters – that act as settings or configuration options for the various methodologies. For instance, in certain analyses I have required that samples contain at least 30 seeds; but the analysis could be repeated with a higher quorum for this parameter – say, 50 seeds – for a more rigorous approach. These key parameters are summarised in Appendix 1. In addition, several of my working assumptions are encapsulated in key bodies of metadata, such as the standardised terminology that I have applied to particular feature types. It would be possible to repeat the analyses undertaken here with different sets of metadata, resulting from different judgements or research aims. I have presented my sets of key metadata in Appendix 2, for reference and for repeatability's sake.

This kind of exposition is intended not only to support replication of the work, with the potential to tweak key variables, but also to admit a rigorous degree of critical assessment by the reader, which is seldom possible with heavily summarised methodologies.

Why crops and weeds?

Prior archaeobotanical work by the author has begun to illustrate the details of arable growth in terms of cereal crop choices. Charred plant remains from the Upper Thames valley and East Anglia, I have argued, demonstrate a diversification in cereals from the 7th century onwards, with cereal-cropping decisions adjusted to best suit local environmental conditions, such that drought-resistant rye was increasingly favoured on sandier soils, for instance, while salt-tolerant barley was preferred in saline regions (McKerracher 2016, and see Chapter 5 of this volume). Such work has, however, only scratched the surface of agricultural ecology in the Mid Saxon period. It assumes that certain cereals were favoured in certain regions because of their general suitability for different terrains, but it sheds little light on how those crops were cultivated in their respective regions.

A far more sensitive proxy for arable growing conditions, and thus for crop husbandry strategies, is offered by the accidentally harvested and charred seeds of arable weeds preserved amongst the cereal grains, since weed floras respond in distinctive ways to environmental variables, and these responses can be studied in modern farmlands managed in traditional ways. It is possible that the characteristics of individual weed species have subtly changed over time and between regions, such that modern ecological observations of a single weed taxon may be a misleading guide to past arable environments. However, this potential difficulty can be overcome by considering a variety of species at once, since it is unlikely that entire weed floras have changed in exactly the same ways since antiquity (Jones 1992: 136–137).

Before analysing the relative economic and ecological significance of different crops and weeds in the archaeobotanical record, however, we must grapple with the complexity of charred crop deposits. Put simply, we must treat deposits of charred botanical remains as artefacts – their charred nature indicating their origins in human activity – and thus assess their taphonomy before analysing their contents. Again, these kinds of assessments can be performed through the application of standardised, repeatable, quantitative methods: for instance, the determination of which stage in the cereal processing sequence is represented by a particular deposit, from the proportions of grain, chaff and weed seed in its contents (see Chapter 4).

Setting the scene

It is not sufficient, of course, simply to lay out a set of methods and present quantitative results. The analyst needs to interpret these results within a heuristic framework based, in this case, around the concept of agricultural development, with clearly defined terminology to avoid inter-disciplinary confusion and ambiguity. Equally, however, agricultural innovations cannot operate in a social vacuum. Before outlining the heuristic framework of this study, therefore, I will first sketch out brief environmental, social and economic narratives of Early and Mid Saxon history, as they pertain to the development of farming practices.

It is long since the Early Saxon period was thought to have witnessed the fruitful genesis of English agriculture. Gone are the days when historians envisaged the heroic salvation of a post-Roman wilderness: forsaken farmland, riven with swathes of regenerating woodland, awaiting the steady hand of the early English yeoman to carve out a new landscape of open fields. Palynological studies have suggested that much of the British landscape was already open farmland by the end of the Roman period, and that in many areas (especially lowland regions) this open farmland persisted throughout the Early and Mid Saxon periods without any large-scale woodland regeneration (Dark 2000: 150–154). The extensive palynological synthesis conducted in the *Fields of Britannia* project identified 'relatively little overall change during the first millennium AD', with no abrupt dislocation at the end of the Roman period (Rippon *et al.* 2015: 312).

The identification of localised woodland regeneration in some more northerly and upland pollen sequences, particularly those in the region around Hadrian's Wall, has however lent support to a new model of agricultural change in the 5th and 6th centuries, in which the most influential factor was not the arrival of Germanic settlers but the collapse of Romano-British economic and administrative infrastructures. Neither the means (villa estates), nor the markets (urban centres and military garrisons), nor the exchange mechanisms (coinage, taxation, state-sponsored transport by land and sea) survived to maintain the large-scale agrarian output that had been required and sustained in Roman Britain. As a result, the post-Roman agricultural economy entered a period of 'abatement', characterised by a shift away from high levels of arable productivity in favour of a greater emphasis on pastoralism (Faith 2009: 24–26).

This is not to suggest that Early Saxon agricultural practices were stagnating and entirely devoid of innovation. Indeed, the apparent shift in emphasis towards pastoralism could be considered an active adaptation to changing circumstances, albeit one which did not entail enhanced productivity *sensu stricto*, since pastoral farming generally produces lower calorific returns per land-unit, in comparison with arable farming (Spedding *et al.* 1981: 355). There are also likely to have been various minor modifications in farming practices throughout this period, the inevitable 'micro-inventions' discussed by van der Veen, implemented by individual farmers as needs and opportunities arose (van der Veen 2010: 7).

It is only from around the 7th century onwards that evidence begins to indicate plausible conditions for, and possible causes of, increasing agricultural productivity. Climatic change may be significant in this regard, but it is not currently understood in sufficient detail to be cited as a factor in – still less a determinant of – agricultural change in this period. A range of evidence, including historical sources, ice cores and tree rings, suggests that cooler, wetter conditions prevailed from around the 5th century onwards, with warmer and drier conditions returning in the last quarter of the millennium, continuing (with much geographical and chronological variability) towards the so-called 'Medieval Warm Period' (Dark 2000: 19-28; Hughes and Diaz 1994). It has further been suggested that the rapidity of climatic change in the 5th and 6th centuries would have been so disruptive as to have had adverse effects on agricultural production (Büntgen et al. 2011: 578-582). However, since such climatic studies lack close chronological and geographical precision, it would be unwise to presuppose any specific, direct relationship between macroclimatic changes and agricultural conditions at a given place and time.

Demographic pressure has long been posited as a basic (though not necessarily sufficient) causal factor behind agricultural development and, although difficult to demonstrate conclusively, could be inferred from a general expansion of settlement patterns in the Mid Saxon period (Hamerow 1999: 417; Morrison 1994: 118-124). Besides a growing population, demands for greater agricultural surpluses could also have come from élites, demanding tribute or renders, and markets, requiring goods for trade and craft-production (Morrison 1994: 125). Both of these can be identified as potential stimuli in the Mid Saxon period. The rise of a new élite in the late 6th and early 7th centuries, with evident command over labour and raw materials, is suggested by the occurrence of rich 'princely' burials and high-status settlements with large timber halls at around this time (Ulmschneider 2011: 159-160; Welch 2011: 269-275). Specific demand for agricultural produce in the form of food-renders, associated with the stabilisation of political structures, is further attested by documentary sources such as the late seventh-century laws of Ine of Wessex (EHD no.32, 70.1). It has been suggested that monastic landlords, documented from the late 7th century onwards in surviving charters, may have exerted particularly strong pressure on Mid Saxon agricultural land in order to produce or procure special ecclesiastical goods, such as vellum for monastic scriptoria (Blair 2005: 251-261).

Another potential stimulus that could be identified in Mid Saxon England – namely, demand for marketable surpluses to support specialist craft and trade activity – is represented principally by the so-called wics or emporia: large, organized settlements such as Ipswich (Suffolk), with populations participating in long-distance exchange and craft production (Cowie 2001: 17). The non-agrarian populations of such settlements are thought to have depended upon surplus goods from rural producers, received via royal food-renders and/or goods-exchange. The possibility that farming communities utilised greater surpluses in order to participate in wider trade networks is raised by the widespread rural distribution of artefacts such as silver sceatta coinage, Rhenish lava querns and Ipswich Ware pottery throughout much of Mid Saxon England, extending beyond the immediate hinterlands of the wics (Blinkhorn 2012: 87-99; Hamerow 2007: 225-226). The so-called 'productive sites', concentrations of coinage and other metalwork, often identified by metal-detectorists, may represent rural markets within this Mid Saxon exchange network (Ulmschneider 2000: 100-104).

The impact of these new markets and élites upon Mid Saxon farming may have extended beyond their demands for greater surpluses: they might also have been directly conducive to agricultural development. For instance, the evidence for long-distance exchange implies intensified inter-regional contacts both within and beyond Mid Saxon England, suggesting a plausible historical context for the diffusion of new technologies, agricultural or otherwise (Ruttan 1998: 158-159). The implementation of new technologies might also have depended upon the existence of supporting institutions (Ruttan 1998: 161–162), represented in Mid Saxon England by the growing power of secular and ecclesiastical landlords. The stimulating influence of strong landlords may have been further enhanced by the introduction of 'bookland': estates held in perpetuity and recorded as such in charters. Monastic examples survive from the late 7th century onwards, secular cases from the late 8th century onwards. The formal longevity of these landholdings could well have encouraged long-term investments in productivity, although it should be noted that, by definition, we are poorly informed as to how bookland differed from other, undocumented forms of landholdings (Blair 2005: 85, 129; Faith 1997: 159–161).

The apparent reorganization of landholdings in Mid Saxon England is a controversial subject. One influential interpretation has been offered by Brown and Foard, who argue from field surveys and excavated evidence in Northamptonshire that settlement nucleation had occurred by c. 850, as part of a process of 'manorialization' whereby peasant cultivators lost their freedom to increasingly powerful landlords (Brown and Foard 1998: 91). Although it is not the main subject of their discussion, they thus imply a model of agricultural development in which landowning élites were directly, purposefully instrumental in the reorganization of the productive landscape. The idea that the nucleated village may ultimately have Mid Saxon origins is reinforced by Reynolds' observation that rectilinear features – 'suggestive of imposed spatial regulation' – appear to be an innovation at Anglo-Saxon rural settlements from the late 6th or early 7th century onwards, and by Hamerow's related argument that Mid and Late Saxon settlements were often more stable than their more mobile, dispersed, and unenclosed Early Saxon predecessors (Hamerow 2012: 67–73; Reynolds 2003: 10–119). The nucleation and stabilisation of rural settlements can be seen as directly conducive to agricultural development, since agricultural labour and technologies may be more effectively deployed if stable and centralised (Williamson 2003: 67–68, 157).

Building a framework

Such is the historical context, in brief, within which agricultural development is understood to have progressed. We must now return to the question of a heuristic framework for exploring the transformation of farming practices, a framework that may ideally be applied to questions of agricultural development regardless of historical context.

Agricultural development exists as a distinct field in economics, in which guise it is concerned primarily with models for the enhancement of modern agrarian productivity in developing countries. A detailed exploration of this field is beyond the scope of this book, but certain concepts may usefully be borrowed to help frame new and existing archaeological hypotheses. So, for a working definition of the term 'agricultural development', I have followed Norton and Alwang's (1993: 170) description of it as the process whereby agrarian productivity can be increased through the stimulation of 'the basic sources of growth (labour, natural resources, capital, increases in scale or specialisation, improved efficiency, and technological progress)'.

Agricultural development, following Norton and Alwang, comprises a series of enhancements which stimulate growth. These enhancements I will gloss as agricultural innovations, each of which entails the transformation of techniques, practices, tools and materials. Evenson (1974: 52) has classified these entities in terms of five technologies – closely interrelated and sometimes overlapping – which form a useful framework for discussing innovations in Anglo-Saxon crop husbandry:

  i. crop-biological,
  ii. animal-biological,
  iii. chemical,
  iv. mechanical, and
  v. managerial.

Under these headings, I will now summarise current ideas about Mid Saxon developments in arable farming which are apt to be investigated by archaeobotanical means. This book's specific focus on arable farming means perforce that animal-biological innovations will be omitted here, but these have been subject to extensive expert consideration in other recent volumes (Crabtree 2012; Holmes 2014).

Crop-biological innovations

Crop-biological innovations are concerned with the range and relative importance of the crop species cultivated. The introduction or reintroduction of certain crop species can be considered a productive innovation if the crops in question are potentially higher-yielding, and/or of greater cultural or economic value, than those previously cultivated. Similarly, there might be shifts in emphasis within the existing crop spectrum, towards crop taxa of greater cultural or economic value.

Archaeobotanical data have been drawn upon in studies of Anglo-Saxon arable farming for many years now, but extensive, data-rich, specialist syntheses have been slow to emerge, and misunderstandings sometimes appear in non-specialist discussions: for instance, *Triticum aestivum* (i.e. bread wheat) being dubbed 'einkorn' (Fowler 1999: 22), or charred henbane seeds being mistaken for pollen (Oosthuizen 2013: 66). Such misunderstandings are of course forgivable but nonetheless significant: the differences between einkorn and bread wheat, and between pollen and charred seeds, are critical in bioarchaeology.

A statistical review led by van der Veen in 2013 has highlighted the strong potential, and vital need, for more intensive specialist research on British plant remains from across the medieval period (van der Veen *et al.* 2013). Nonetheless, several crop-biological innovations have already been identified by those researching the history and archaeology of agricultural change in Early and Mid Saxon England. The most important of these models concern the role of free-threshing wheat, normally understood to be bread wheat (*Triticum aestivum* L.), destined to become the dominant cereal crop of modern Britain. According to most studies, free-threshing wheat supplanted first spelt as the most important wheat crop of Anglo-Saxon England, and then hulled barley as the most important cereal crop overall (Banham 2010: 179; Hamerow 2012: 146; Moffett 2011: 348–351).

It is widely accepted that, whereas spelt (*Triticum spelta* L.) was the predominant wheat crop of Roman Britain, free-threshing wheat came to replace it in the Anglo-Saxon period, but it remains an open question as to when, where, why, and how rapidly this change occurred (Green 1981: 133; Moffett 2011: 349). Some have dated this shift to around the 8th century (Astill 1997: 199; Oosthuizen 2013: 64) but it is more often suggested, more or less implicitly, that the demise of spelt occurred fairly rapidly around the 5th century. Hence, spelt remains found within Early or Mid Saxon contexts are often taken to represent residual prehistoric or Roman activity (e.g. Fryer in Atkins and Connor 2010: 102;

cf. Pelling 2003: 103). An alternative interpretation of spelt remains preserved in post-Roman contexts where residuality or disturbance is considered unlikely, is that spelt may have persisted as a self-seeding volunteer – or, at best, a very minor crop – in fields continuously cultivated since the Roman period (Murphy 1994: 37).

Emmer wheat (*Triticum dicoccum* Schübl.) existed as a minor crop in Roman Britain, and like spelt it has seldom been considered a significant member of the Anglo-Saxon crop spectrum. However, Pelling and Robinson have argued that emmer was reintroduced between the 5th and 9th centuries, locally to the Upper and Middle Thames valley, as part of an agricultural tradition imported by Anglo-Saxon colonists. Their principal supporting evidence comprises glume bases from Dorney (Bucks), radiocarbon-dated to cal. AD 435–663, and from Yarnton (Oxon), radiocarbon-dated to cal. AD 670–900. Without such radiocarbon determinations, a post-Roman date for these emmer macrofossils might have been seriously questioned, given the widespread assumption that Anglo-Saxons did not grow this crop (Pelling and Robinson 2000).

In a later study, drawing extensively upon archaeobotanical evidence from the Upper and Middle Thames valley in general and Yarnton in particular, emmer has been cited along with rye, lentil, grape and plum as newly reintroduced crops of the Mid Saxon period. According to this model, these crop-biological innovations were part of a process of 'agricultural recovery' in the Mid Saxon period, entailing the cultivation of a wider range of crops, agricultural and horticultural, than that evidenced for the Early Saxon period. The authors thus posit innovation through diversification (Booth *et al.* 2007: 329–336), an idea that recurs in other studies. The cultivation of rye (*Secale cereale* L.) and oats (*Avena sativa* L.), in particular, is often thought to have grown in importance over the course of the Anglo-Saxon period (Banham 2010: 179; Hamerow 2012: 150). Diversification may also have embraced fibre crops such as flax and hemp, although views on the role of these crops in Anglo-Saxon farming have varied. For Oosthuizen, their production 'on an industrial scale' was a Mid Saxon innovation; whereas the authors of *Thames Through Time* simply state that flax 'remained common in waterlogged deposits' in both periods (Booth *et al.* 2007: 330; Oosthuizen 2013: 64).

In another model, developed primarily by Banham but with wider currency in Anglo-Saxon scholarship, one specific crop-biological innovation is held to characterise Anglo-Saxon agricultural development: the rise of bread wheat to become the dominant cereal crop in place of hulled barley (*Hordeum vulgare* L.). This model I have elsewhere termed the 'bread wheat thesis' (McKerracher 2016). It is related to, but distinct from, the observed demise of spelt wheat in favour of free-threshing varieties. It focuses upon the relationship between barley and bread wheat, and argues that barley was the most important cereal crop of the Early Saxon period, but that bread wheat rose to dominance from the Mid Saxon period onwards (Banham 2004: 13–14; Fowler 1981: 279; Hagen 2006: 33–35; Hamerow 2007: 225; Oosthuizen 2013: 64).

This model has its origins in Jessen and Helbaek's landmark study of ceramic grain impressions (Jessen and Helbaek 1944), and found renewed relevance with the blossoming of British archaeobotany in the late 1970s. The findings of Monk, in particular, lent weight to the idea of a bread wheat ascendancy (Monk 1977: 332–340). Working with an even larger archaeobotanical dataset, of national scope, Banham found that barley was more common than wheat in the Early Saxon period, but that wheat was more common that barley by the Mid Saxon period – trends which echoed those previously observed by Monk (Banham 1990: 38). Since most of this wheat was positively identified as free-threshing, bread-type wheat, Banham suggested that a dietary preference for wheat bread over barley bread was ultimately responsible for the rise of the former over the latter. Indeed, there are persuasive indications of wheaten bread's higher esteem in the slim documentary record for the period (Banham 2010: 179). Other possible advantages of bread wheat include ease of processing (it is a free-threshing cereal and therefore does not demand the heavy additional labour of dehusking), its winter-hardiness, and the potentially great yields achievable with intensive manuring. However, as Moffett has argued, none of these factors is incontrovertible or necessarily exclusive to bread wheat; and none alone would necessarily explain why free-threshing wheats, although known in Britain since the Neolithic, only achieved predominance from the Anglo-Saxon period onwards (Moffett 2011: 349–350). Thus, as Banham argues, dietary preferences may have been a more decisive factor than purely practical considerations in shaping the crop-biological innovations of this period.

Chemical innovations

Chemical innovations concern the edaphic conditions of arable production, and could have entailed such techniques as crop rotation, fallowing, and the use of heavier, more fertile soils. A Mid Saxon shift in settlement patterns, involving the reoccupation of heavy clay soils for the first time since the Roman period, has long been recognized from both excavated evidence and field surveys, for example in the Sandlings area of Suffolk (Arnold and Wardle 1981; Hodges 1989: 62; Newman 1992: 30–35). That these soils were also tilled rather than just settled and grazed has been inferred from the increased occurrence, at sites such as Yarnton, of stinking chamomile seeds (*Anthemis cotula* L.) amongst Mid Saxon crop remains, since this is a weed characteristic of heavy clay soils (Stevens in Hey 2004: 362).

Zooarchaeological studies, meanwhile, have shown that cattle and sheep were increasingly being kept to a greater age in the Mid Saxon period (Crabtree 2012). This could have made available increasing amounts of animal manure for the intensified enrichment of arable soils. In addition, indirect evidence of Mid Saxon middening can be found in the occurrence of weed species such as henbane (*Hyoscyamus niger* L.), which is characteristic of middens, again in crop assemblages at Yarnton (Hey 2004: 48–49). It should be noted, however, that henbane does not grow exclusively around middens: it is a nitrophilous species that might equally thrive in other 'disturbed and enriched' soils, such as those alongside bridleways or droveways (Cappers and Neef 2012: 111).

Weed ecology therefore provides a sensitive way of exploring chemical innovations but, with a few exceptions such as the work at Yarnton, this approach has yet to be applied to many archaeobotanical assemblages of Anglo-Saxon date. In addition, it is preferable to consider weed floras in their entirety, rather than relying upon individual species as indicators of environmental characteristics, since the latter approach is more vulnerable to bias through chance occurrences and diachronic changes in individual species' ecological attributes (Jones 1992: 136–137).

Mechanical innovations

Chronologically speaking, Anglo-Saxon agriculture falls well within the era of so-called 'pre-mechanised' farming, but there is one particular mechanical innovation that has long been central in the historiography of early medieval arable: the heavy plough. The heavy mouldboard plough is distinguished from the lighter 'ard', or scratch plough, by the addition of a coulter to slice the soil vertically and a mouldboard to turn the sod (Bowen 1961: 7–11).

Artistic, documentary, artefactual and plough-mark evidence for heavy ploughing has long been known from the Late Saxon period, and a form of heavy plough, if not a true mouldboard plough, seems to have existed in Roman Britain (Booth *et al.* 2007: 288; Fowler 2002: 152–153; Hill 2000: 11–13; Williamson 2003: 120). An iron coulter from a swivel plough, a variety of mouldboard plough, has been discovered in an early seventh-century context at the royal complex of Lyminge, showing at least that the technology was available to Kentish kings at this early date (Thomas *et al.* 2016). What remains unclear is how widespread and significant this mechanical innovation was in arable farming at large in Mid Saxon England. The reoccupation of heavy clay soils from the 7th century onwards, as noted above, could be taken to imply the adoption of mouldboard ploughing, since according to Williamson these soils could not have been extensively tilled without a heavy plough (Williamson 2003: 121). On the other hand, with the growth of early medieval settlement archaeology, the picture of a Mid Saxon shift towards the occupation of heavy soils is by no means as clear as it was in the 1980s (McKerracher 2018: 34-37); and in any case the location of a settlement is not in itself a definitive guide to its patterns of land-use.

The weed spectra in the charred crop assemblages at Yarnton have been used as a more direct proxy for heavy ploughing, since species which are more tolerant of soil-disturbance (mostly annuals) are seen to become more prevalent in the Mid Saxon period, while those which are less tolerant of disturbance (mostly perennials) decline (Stevens in Hey 2004: 363-364). More specifically, Martin Jones has argued that a Mid Saxon decline in common chickweed (*Stellaria media* (L.) Vill.) and bromes (*Bromus* L.) in the archaeobotanical record, and a concomitant rise in stinking chamomile (*Anthemis cotula* L.) and cornflower (*Centaurea cyanus* L.) might reflect a shift from 'shallow' to 'deep' cultivation and so, by implication, from ards to heavy ploughs at Pennyland (Bucks). This interpretation is based partly upon the longer seed-dormancy of stinking chamomile and cornflower, and the particular capability of their seeds to germinate in disturbed soils (Jones in Williams 1993: 173-174; Jones 2009).

Managerial innovations

Managerial innovation can embrace a variety of changes concerned with the overall planning and rationale of agricultural production, and one example that has been postulated for the Mid Saxon period is the expansion of arable production at the expense of pastoral farming, as suggested by the appearance of larger-scale facilities for the processing and storage of crops. Specialist grain storage facilities do seem to be an innovation of this period, likewise watermills and grain ovens (Hamerow 2012: 151-152; McKerracher 2014b; Watts 2002: 72-82).

Arable expansion has otherwise been deduced from palaeoenvironmental evidence, especially from pollen sequences in which proportions of cereal pollen increase notably from the Mid Saxon period onwards; and also, at Yarnton, from the greater abundance of cereal remains among charred macrofossils (Booth *et al.* 2007: 333; Rippon 2010: 58). Palaeohydrological evidence is also suggestive of arable expansion beginning in the later Mid Saxon period: alluviation, presumed to reflect soil-erosion consequent on the extension of ploughed land, is seen to increase from around the 9th century onwards in the Nene and Upper Thames valleys (Booth *et al.* 2007: 19-20; Brown and Foard 1998: 81-82; Robinson 1992: 205-206).

Most fundamentally, whole new agricultural systems may have been instituted in this period. For example, a regime known as 'convertible husbandry', which entailed long-term rotations between cereal and grass crops, has been identified as an innovation of the 7th to 9th centuries in a palynological study in Devon (Rippon *et al.* 2006: 55). Oosthuizen meanwhile, through a seminal study of landscape archaeology, documentary sources, and the wider historical context, has dated a 'proto-open-field' arrangement in the Bourn Valley (Cambs) to the Mid Saxon period (Oosthuizen 2005: 176-188).

Additionally, Banham's bread wheat thesis might have further implications in terms of managerial innovation. Banham proposes that bread-type wheat would largely have been autumn-sown, whereas barley would have been spring-sown, as in regimes documented later in the medieval period. In this case, an increase in bread wheat cultivation in the Mid Saxon period could have entailed a shift in the seasonality of sowing regimes, with a greater importance being attached to autumn sowing: 'If changing from barley to wheat meant growing winter corn for the first time, improved drainage might be vital to prevent the young plants standing with their feet in water over the winter, even on soils which were not particularly wet in the spring and summer' (Banham 2010: 183). In order to facilitate the survival and success of bread wheat in those wet winter soils, better drainage could have been afforded by heavy ploughing and its resultant ridge-and-furrow patterns, perhaps culminating in the development of open field systems (Banham 2010: 182–187).

Summary

Such are the innovations that have been thought, in various recent studies, to characterise Mid Saxon farming. More specifically, I have described those possible innovations whose impact may be detectable in the archaeobotanical record. The various ideas discussed above can be distilled into three broader themes, around which the remainder of this study is based. Hence, the objective of this book is to determine how, when and where:

  i. crop surpluses grew,
  ii. crop spectra shifted, and
  iii. crop husbandry regimes changed, in Early and Mid Saxon England.

The following five chapters will provide worked examples of how descriptive, quantitative and semi-quantitative archaeobotanical analyses can produce results directly relevant to these themes. Chapter 2 provides a technical and descriptive account of the dataset that underlies this book's analyses, and Chapter 3 discusses the plant taxa which constitute that dataset, and how they are treated in this study. Chapter 4 addresses the question of crop surpluses, by considering the character, distribution, abundance and density of charred crop deposits in the project dataset. Chapter 5 investigates changes in crop spectra through the combined application of semi-quantitative and fully quantitative analyses of charred cereal remains. Chapter 6 turns to the evidence of charred weed seeds, a powerful proxy for changes in crop husbandry strategies and arable environmental conditions. Finally, in the closing chapter I will revisit the themes and theories discussed above, and consider what the findings of this study may contribute to those debates.

Chapter 2: Describing the Data

The amount of archaeobotanical data available for early medieval England is very large and continuously growing. It constitutes a highly complex dataset, derived from the work of many different archaeobotanists, conducted over several decades. Each analyst has brought different experiences, reference materials, idiosyncrasies, budgets and time constraints to their work. The original materials behind their data – that is, the grains, chaff, seeds and other botanical items – vary considerably both in quantity and in quality of preservation, within and between different assemblages. Methods of excavation, local environmental conditions and soil sampling strategies at different sites are also diverse, and likely to have exerted important influences on archaeobotanical results. Charred remains dominate the British archaeobotanical record, but material is sometimes preserved in other ways too: by concealment within anoxic waterlogged environments, by mineral-replacement in phosphate-rich deposits, or by impression in fired clay (the botanical equivalent of dog pawprints on Roman tiles).

All of these complicating factors necessitate a clear and precise description of the data being used in an archaeobotanical study, so that the reader may understand, as far as is practical, the bases of inference and analysis. Such is the purpose of this chapter.

Regional scope

Given the potential influence of regional variability upon patterns of agricultural development in Early and Mid Saxon England, I reasoned that this investigation would benefit from a comparative approach, drawing upon evidence from two regions which encompass different landscapes and environments. A fair comparison requires that those regions share a similar abundance of archaeological evidence for the Early and Mid Saxon periods. Prior investigation suggested that the following regions, arbitrarily defined as units of manageable size, would be suitable as case studies: one region centred around the Upper and Middle Thames valley and environs; and another encompassing East Anglia and Essex (Figure 1). The intervening counties of Northamptonshire, Bedfordshire, Hertfordshire and Greater London were omitted only because of the project's inevitable time constraints.

As intended, the selected regions incorporate considerable topographical variety, including, for example, inland and coastal landscapes, chalk downs and river gravels, hills and fens. Moreover, excavations across the two regions have discovered relatively large numbers of Anglo-Saxon settlements. As the corpus of known sites continues to grow through development-led excavations, the archaeological authenticity of an Anglo-Saxon culture zone in southern Britain centred on the Great Ouse and Upper Thames valleys has become increasingly convincing (Blair 2013: 5–8; Hamerow 2011: 120; Hamerow 2012: 4, Fig. 1.1).

The boundaries of the two case study regions were defined in terms of post-1970 administrative counties and unitary authorities, in order to facilitate research via the Historic Environment Records maintained by local government bodies in England. Hence these two zones do not necessarily have any intrinsic environmental or historical significance. For analytical purposes, therefore, different geographical divisions have been employed, in a scheme more sensitive to landscape variation: National Character Areas, as discussed further below.

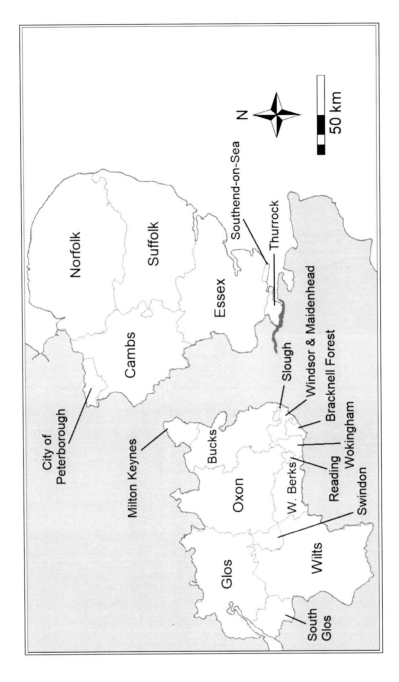

Figure 1 - Modern administrative geography of the case study regions.

Data collection was initiated through the consultation of resource assessments, research agendas, fieldwork summaries in journals, and online databases such as Pastscape (<https://www.pastscape.org.uk>, accessed January 2019). From these sources, excavated sites with Early and Mid Saxon settlement remains were identified. References were then pursued in published and unpublished sources and, where archaeobotanical data were available, these were recorded in Microsoft Excel spreadsheets, along with basic metadata about each site. In several instances, where relevant data were unpublished, or published in insufficient detail, archive reports were sought directly from excavators, environmental analysts, or local authorities. An arbitrary threshold was set such that excavation reports predating 1970 were not utilised, since before this date many excavations will not have been conducted to modern scientific standards, nor will bioarchaeological remains have been systematically recovered from Anglo-Saxon settlements (Fowler 1976: 46–47).

The resulting collection of data – the foundation of this book – cannot be considered complete or truly comprehensive. Samples from graves and 'off-site' bioarchaeological data, e.g. from isolated palaeochannels, were deliberately omitted from the dataset, partly because of their less straightforward relationship to human settlement activity, but also to place practical limits on the extent of the investigation. It should also be noted that data collection ceased in the spring of 2012, when further trawling was yielding rapidly diminishing returns, and important new data will doubtless have emerged during the intervening years. For the present volume, I have updated some bibliographical references but the dataset remains otherwise unchanged.

Geographical locations were recorded using National Grid References (NGRs), so that spatial patterns in the data could be explored using a Geographic Information System (GIS). If not explicitly stated in an excavation report, NGRs were obtained from historic environment databases or read from site plans and maps. Such NGR locations are normally given in the form of a grid square followed by x and y coordinates within that grid square, in hundreds of metres: for instance, 'SP 460 320'. These I have translated to eastings and northings, centred to the nearest five metres, for mapping purposes, and in this form are they detailed in Appendix 3, a gazetteer of all sites included in the project dataset.

The geological and topographical situations of each site were studied because, as determinants of local soil composition, they are likely to have affected both archaeological preservation conditions and past agricultural environments (Goldberg and Macphail 2006: 60). Where geological information was not available in excavation reports, the British Geological Survey's online Geology Roam facility, accessed via the EDINA Digimap service, was consulted (<http://digimap.edina.ac.uk>, accessed January 2019). Such environmental information is too extensive to reproduce in this book, but brief summaries for the sites in the project dataset can be found in the appendix to the companion volume, *Farming Transformed* (McKerracher 2018).

While local geology and topography are important, however, a settlement's agricultural activities are likely to have extended beyond its immediate vicinity and into other terrains, if only occasionally or seasonally. To allow for this possibility, site distributions were also considered within the context of National Character Areas (NCAs), a framework devised in the 1990s by the government body English Nature (now Natural England). Descriptive profiles and GIS data pertaining to the NCAs have been updated and published over the last ten years (<https://www.gov.uk/government/publications/national-character-area-profiles-data-for-local-decision-making>, accessed January 2019). NCAs define areas with similar landscape characteristics – like their precursors, the Natural Areas – constituting 'biogeographic zones that reflect the geological foundation, the natural systems and processes and the wildlife in different parts of England,' which transcend the boundaries of modern administrative units (Webster 2008: 3).

It should be remembered, however, that NCAs are based upon modern observations and should be treated as such: past environments need not be expected to correspond exactly with those described by Natural England, although certain geological attributes may have remained constant. On the other hand, unlike descriptive accounts of individual sites' topographical and geological attributes, the use of open-access GIS data and predefined landscape zones is consistent with a standardised, repeatable approach to analysis, which is a cornerstone of this study. For this reason, National Character Areas form the basis of geographical analyses in this book. Specifically, the following NCAs are represented by the sites in the project dataset (Figure 2). The identification numbers are those given by Natural England.

- **46 – The Fens:** low, flat, extensive wetlands, including the saline, marshy silt fens by the Wash (i.e. nearest the coast) and, further inland, the peaty Black Fens with islands of sandstone, sand and gravel.
- **76 – North West Norfolk:** open landscape with light, fertile soils.
- **79 – North East Norfolk and Flegg:** coastal grasslands, scrub and dunes.
- **81 – Greater Thames Estuary:** low estuarine terrain with mudflats and saltmarsh.
- **82 – Suffolk Coast and Heaths:** low-lying terrain with low rainfall and dry, acid, sandy soils, encompassing an area known as the Sandlings.
- **83 – South Norfolk and High Suffolk Claylands:** boulder clay plateau, dissected by rivers cutting the underlying chalk.
- **84 – Mid Norfolk:** plain of slow-draining but sometimes fertile chalky boulder clay, with sands and gravels in river valleys.
- **85 – Breckland, or 'Brecks':** chalk plateau largely covered by dry, acid, sandy soils.
- **86 – South Suffolk and North Essex Claylands:** fairly fertile soils formed on chalky boulder clay, cut by small river valleys.
- **87 – East Anglian Chalk:** uplands bearing thin calcareous soils, and gravel terraces in the dissecting river valleys.
- **88 – Bedfordshire and Cambridgeshire Claylands:** rolling clayey landscape with sands and gravels in shallow river valleys; some soils are fertile, some damp, and some well-drained.
- **90 – Bedfordshire Greensand Ridge:** sandstone ridge surrounded by the Beds and Cambs Claylands, with heathland and floodplain among its habitats.
- **92 – Rockingham Forest:** limestone ridge, with heavy clay soils.
- **106 – Severn and Avon Vales:** low-lying, variable, riverine terrains.
- **107 – Cotswolds:** limestone uplands bearing thin soils, dissected by river valleys with clays and gravel terraces.
- **108 – Upper Thames Clay Vales:** fertile landscape with heavy clays on the valley sides and lighter gravel terraces.
- **109 – Midvale Ridge:** limestone hills within the Upper Thames valley, bearing sandy, acid soils.
- **110 – Chilterns:** light, calcareous soils in the chalk hills; limestone and clayey deposits at the foot of the escarpment, meeting the Vale of Aylesbury.
- **111 – Northern Thames Basin:** variable soils, from poor, slow-draining terrain on the London Clay, to better soils on riverine alluvial deposits.

- **115 – Thames Valley:** clays and acid sands on the flinty gravel terraces of the Middle Thames valley, a landscape less fertile than the Upper Thames valley.
- **116 – Berkshire and Marlborough Downs:** rolling chalk downland.
- **117 – Avon Vale:** undulating landscape with seasonally-flooded watercourses and wooded upper slopes.
- **129 – Thames Basin Heaths:** clays and acid sands on the flinty gravel terraces of the Middle Thames valley, a landscape less fertile than the Upper Thames valley.
- **132 – Salisbury Plain and West Wiltshire Downs:** chalk uplands with extensive, distinctive calcareous grassland.

Structure of the dataset

The archaeobotanical dataset consists of records, samples and assemblages. The specific definitions of these terms, as used within this project, are outlined below.

An *assemblage* is a collection of archaeobotanical material belonging to a particular phase at a given site, with a common means of preservation. Thus, for example, Yarnton has three charred assemblages of Early, Intermediate, and Mid Saxon date respectively, and one waterlogged assemblage of Mid Saxon date (Hey 2004). Each assemblage consists of one or more samples.

A *sample* is a collection of archaeobotanical material with a common means of preservation, deriving from a common archaeological context. Each sample consists of one or more records. Samples have been assigned codes, comprising letters to denote the parent site and sequential numbering, enclosed within chevrons, e.g. <FLX20>. These codes are specific to this project. The sample inventory in Appendix 4 lists the sample and context numbers used in the original excavation reports, where available, allowing the reader to trace the original data for each sample. Where a single context has been sampled several times, and is clearly represented by multiple 'samples' in its original report, I have amalgamated the data for these samples unless they exhibit differences in botanical composition which may suggest that the material in question does not, in fact, represent a single deposit. For example, my sample <Y34-5> from Yarnton is an amalgamation of two original soil samples which both derive from context 3314, described as a 'layer'. In this way, as far as possible, the term sample has been reserved for plausibly discrete, independent deposits.

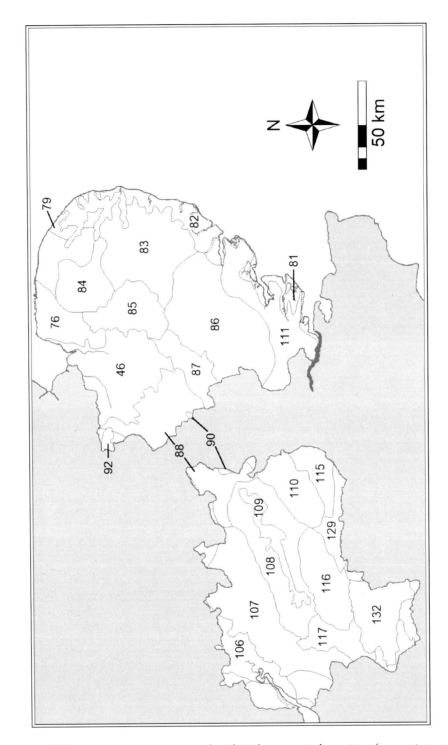

Figure 2 - National Character Areas represented within the case study regions (see main text for key to identification numbers).

Finally, for the purposes of this study, records are the constituent members of samples. Each record is defined by the following items of data:

1. Taxonomic identification of a plant, whether familial, generic or specific (e.g. Poaceae, *Avena*, or *Avena sativa*).
2. Anatomical identification of the plant part (e.g. 'seed').
3. Quantity of items represented (whole items, or equivalent minimum number of individuals).
4. Fragmentation, i.e. fragmented or whole.
5. Modifiers or qualifiers, denoting levels of precision or confidence in the taxonomic identification (e.g. 'cf.' or 'type').

Fragmentation was recorded as a binary variable in each record, to distinguish between whole items and fragment counts. Archaeobotanical material is often fragmented to some extent, but the quantification of plant parts can be standardised by counting selected diagnostic zones (e.g. embryo ends of grains) rather than all individual fragments, and thus calculating minimum numbers of individuals (MNI) for each plant part (Jones 1990: 91–92). Quantification procedures are not always specified in archaeobotanical reports, but it has been assumed that all counts refer to whole items or equivalent MNIs unless fragment counts are explicitly specified. Fragment counts have been excluded from all quantitative analyses but have been considered in semi-quantitative presence analyses.

Independence

As Glynis Jones has argued, archaeobotanical data are most usefully analysed at a level which can be related to individual behavioural or depositional activities (Jones 1991: 64). Hence, it is important to determine whether samples are independent (each representing a different activity), interdependent (several representing a single activity), or composite (each representing several activities). These distinctions are important because the associations between different taxa can only be usefully investigated in independent samples (Bogaard 2004: 61–62). Composite samples, by contrast, may exhibit spurious co-occurrences of taxa that were not in fact grown, processed or deposited together, but whose co-occurrence is entirely due to redeposition or post-depositional factors.

As mentioned above, interdependent samples were amalgamated to produce independent samples, if their interdependence was implied by the available stratigraphic information and consistent with their botanical composition. For the avoidance of doubt, samples whose interdependence was suspected but not demonstrable were excluded from sample-based analyses. Data from composite and interdependent samples were nonetheless eligible for inclusion in assemblage-based analyses. It must be admitted that this sifting and derivation of independent samples is, ultimately, a subjective process that cannot easily be parameterised or reduced to truly replicable procedures. For this reason, it is necessary to make clear how original sample data have been treated in the project dataset: whether omitted, amalgamated, or accepted verbatim as independent. This information can be found in the inventory of samples (Appendix 4).

Preservation

Modes of preservation – e.g. charring or waterlogging – were specified for each record in the dataset. This is a crucial variable, since the archaeobotanical representation of taxa and plant parts can be affected by differential preservation biases: charring, for example, is biased towards crops (and their associated weeds) whose processing sequences involve fire, especially cereals. Waterlogged deposits, by contrast, have potentially more diverse catchments (Dennell 1976a: 231; Green 1982: 42–43).

Reliable comparisons between the occurrences of different taxa, and their different parts, therefore require that we compare like with like, i.e. that the dataset be divided into discrete subsets, each defined by a single mode of preservation.

To this end, each sample was assigned a principal means of preservation, which accounted for at least 70% of all whole plant parts in that sample (Appendix 1: Parameter 1). The subsets of data were then analysed exclusively, i.e. analyses of the charred dataset consider only those samples with charring as their principal means of preservation, and include only the charred records within those samples. Small quantities of differently preserved material (e.g. mineralized seeds in a predominantly charred sample) were omitted as potential contaminants. The same principle applies to assemblage-level analyses: a 'charred assemblage' is taken exclusively to mean the charred records from a set of samples whose principal means of preservation is charring.

A separate classification, ceramic assemblage, was applied to all plant remains preserved as impressions in fired clay. This material cannot constitute 'samples' or 'assemblages' in a sense comparable to the charred, waterlogged or mineralized material. The formation processes of ceramic plant impressions are fundamentally different from those of plant remains derived directly from soil samples: the preservation and identification of these impressions may owe as much to pottery taphonomy as to ancient plant-use (Dennell 1972; 1976b: 13).

In summary, four modes of preservation are represented among the Early and Mid Saxon plant remains from the regions studied in this book: charring, waterlogging, mineral-replacement and ceramic impression. However, an overwhelming majority of the material noted during this study (111 of 137 assemblages) was found to be charred. Since this is not only the best-represented category of archaeobotanical evidence, but also that best suited to the systematic investigation of cereal cultivation and processing, the dataset used in the rest of this book consists entirely of charred plant remains. The term 'project dataset' should therefore be understood, from this point on, exclusively to denote the charred plant remains from 736 charred samples, constituting 111 assemblages at 96 sites across the study regions.

There are some disparities in the nature and quality of these data. For six of the 96 sites with charred remains, data are available only at the level of assemblages, not for individual samples: Great Linford, Stonea Grange, Two Mile Bottom, Wickhams Field, Worton, and Walton Vicarage. Among the 736 individually detailed samples, 578 contain fully quantified items (including one with fragments only, and no whole items or estimated equivalents) while 158 have only semi-quantitative data, i.e. indicating each taxon's presence, or its abundance on a simplified scale, rather than giving the precise quantity of its charred remains. The 15 samples from Wilton are slightly problematic in that only chaff items and non-cereal seeds are individually quantified; for cereal grains, the total quantity in each sample is given, but individual taxa are not separately quantified. Hence, these 15 samples are not suitable for the quantitative analyses employed in this book which, as will become clear, require grains, chaff and weed seeds all to be fully and separately quantified.

Analysts

Prior to the archaeobotanical analysis of the project dataset, the potentially biasing effects of inter-worker variability were considered. Differences in methodology, research design, scheduling, expertise and equipment could all have introduced biases into the dataset that may be mistaken for genuine archaeobotanical trends. While there is no comprehensive way in which such biases can be fully mitigated, a basic assessment of their potential impact can be made through an investigation of analyst representation, i.e. counting the number of assemblages upon which each person has worked. Since some assemblages were worked on by more than one person, the total number of analyses (116) slightly exceeds the total number of assemblages (111).

The dataset represents the work of 29 archaeobotanists (including one anonymous), eight of whom are especially prolific, accounting for more than 70% of all analyses: Martin Jones, John Letts, Rob Scaife, Mark Robinson, Ruth Pelling, Chris Stevens, Peter Murphy and Val Fryer (Table 1; Figure 3). The dominance of these eight analysts lends some degree of inter-worker uniformity to the dataset, although it should be noted that their work is not evenly distributed within the study regions. The work of Fryer and Murphy, for example, is almost exclusively restricted to East Anglia and Essex, while that of Pelling, Robinson and Letts is heavily concentrated in and around the Upper and Middle Thames valley. The work of the 21 less well-represented analysts may be considered to have lower comparative value, although this is certainly not to suggest that their studies are any less reliable.

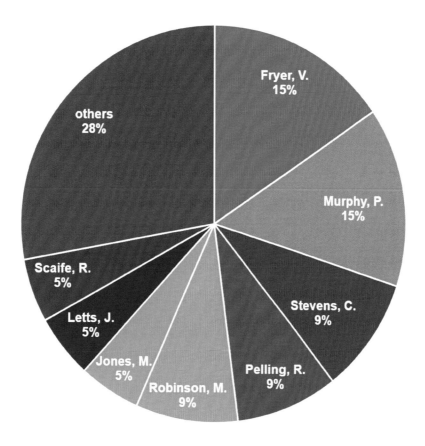

Figure 3 - Proportions of analyses undertaken by different archaeobotanists within the project dataset (percentages rounded to the nearest whole number).

Table 1 – Representation of archaeobotanists in the project dataset.

archaeobotanist (n=29)	no. analyses (n=116)	% analyses
(anonymous)	1	0.9
Ballantyne, R.	3	2.6
Busby, P.	1	0.9
Carruthers, W.J.	3	2.6
Clapham, A.	2	1.7
Cramp, L.	1	0.9
Evans, L.	1	0.9
Fryer, V.	18	15.5
Giorgi, J.	2	1.7
Jones, A.K.G.	3	2.6
Jones, J.	1	0.9
Jones, M.	6	5.2
Letts, J.	6	5.2
Livarda, A.	1	0.9
Martin, G.	1	0.9
Moffett, L.	1	0.9
Monk, M.	1	0.9
Murphy, P.	17	14.7
Nye, S.	1	0.9
Pelling, R.	10	8.6
Roberts, K.	3	2.6
Robinson, J.	1	0.9
Robinson, M.	10	8.6
Scaife, R.	6	5.2
Smith, W.	2	1.7
Stevens, C.	11	9.5
Straker, V.	1	0.9
van der Veen, M.	1	0.9
Vaughan-Williams, A.	1	0.9

Chronology

In the vast majority of cases, the samples in the dataset were not directly dated in their original excavation reports, but were rather assigned to phases within the generalized chronological sequences of individual sites. Clearly this is problematic, since it creates somewhat artificial populations of flora which are time-averaged in terms of centuries, blurring any diachronic developments that may have occurred over shorter timescales. Moreover, the common assumption that bioarchaeological remains are contemporary with stratigraphically associated artefacts may be misleading, especially in the case of tertiary deposits which may represent the accumulation of several depositional events over an extended period (e.g. the backfills of some *Grubenhäuser* according to Tipper 2004: 157–159). Such problems are compounded by the fact that ceramic chronologies for the Anglo-Saxon period are regionally variable and generally of poor definition. For example, while Ipswich Ware is a near-ubiquitous indicator of 8th and 9th century dates in much of East Anglia, parts of the Upper Thames valley may have been essentially aceramic at this time (Blinkhorn in Hardy *et al.* 2003: 172). Even where charred plant remains have been radiocarbon-dated, these determinations frequently span two or more centuries, and dates produced prior to the publication of widely-accepted calibration curves in the 1980s and 1990s might not be reliable (Pollard 2009: 154).

In light of these difficulties and ambiguities, it was decided to assign samples and assemblages, with as much precision as available data allowed, to four broad and overlapping chronological categories based upon scientific dating results (especially radiocarbon) and stratigraphic associations with artefacts and settlement features. The intention is to allow at least a relative chronology – an ordinal progression – so that possible developmental sequences may be examined, even if the absolute dating of particular developments remains imprecise. It is based upon the conventional division between Early (*c.* 410–650) and Mid Saxon (*c.* 650–850) phases, but attempts a slighter greater degree of precision by accounting for Blinkhorn's refined chronology for Ipswich Ware (Blinkhorn 2012) and observations by Hamerow and Reynolds on developments in settlement morphology (Hamerow 2012: 67–72; Reynolds 2003: 110–119). It thus attempts to define Early Saxon (*c.* 410–650), Intermediate (*c.* 600–850), Mid Saxon (*c.* 720–850+), and Generic (*c.* 410–850) phases.

Early Saxon assemblages are those deemed largely to pre-date *c.* 600, although they may include some early seventh-century material. Besides radiocarbon determinations falling principally within the 5th and 6th centuries, Early Saxon dates are chiefly characterised by (i) ceramic assemblages with relatively high proportions of decorated handmade sherds and/or relatively low proportions of organic-tempered sherds; (ii) rim sherds whose morphology is considered to be of fifth- or sixth-century date; (iii) annular loomweights; (iv) other chronologically diagnostic artefacts such as ornamental metalwork; and (v) stratigraphic relationships with well-defined Late Roman and Intermediate/Mid Saxon phases.

Intermediate assemblages are those deemed largely to post-date *c.* 600, although they may conceivably include some late sixth-century material and, in some cases, could theoretically be dated as late as the 9th century. Hence, this material could potentially have considerable chronological overlap with Mid Saxon assemblages, which are deemed largely to post-date *c.* 720 but may include some material of later seventh-century date and could, in some cases, extend into the later 9th century.

The Intermediate horizon (post-600) is characterised chiefly by (i) relatively high proportions of organic-tempered sherds among handmade ceramics, with Ipswich Ware and decorated sherds very rare or absent, although Maxey-type ware may be present in some areas; (ii) the use of linear ditch systems within settlements; (iii) beamslot or post-in-trench construction; (iv) axial arrangement of buildings; (v) larger *Grubenhäuser*; (vi) coinage of seventh-century date (e.g. Primary sceattas); and (vii) stratigraphic relationships with well-defined Early and/or Mid Saxon material. The Mid Saxon horizon (post-720) can be characterised by any of the Intermediate qualifiers listed above, plus coinage of eighth- to ninth-century date (e.g. Secondary sceattas and Mid Saxon pennies), and Ipswich Ware, especially in East Anglia. The distinction between Early, Intermediate and Mid Saxon phases is most secure when they occur at a single site. Yarnton and Pennyland, for example, both have material belonging to all three phases.

The object of this chronological approach is to determine a likely *terminus post quem* for any given agricultural innovation that is apparent in the data, i.e. whether it only demonstrably occurs from the 8th century onwards, or whether it could already be apparent from the 7th century. Given that the distinctions between the phases can sometimes be blurred, such inferences should be considered probabilistic rather than definitive.

It was often impossible to obtain chronological precision beyond a span of four or more centuries, e.g. 5th to 8th, or 6th to 9th centuries, usually because of a lack of diagnostic pottery types. Such material is here assigned to a Generic category. It should be understood that material of Generic date could derive from any sub-period between the 5th and 9th centuries, and does not necessarily span the entire 5th to 9th century period. Although not useful for exploring chronological patterns, Generic material can nonetheless be utilised in geographical analyses.

The assignment of data to these chronological categories is subjective, and it is readily acknowledged that the dates used in this project may well be disputed by other scholars, or require revision in the light of new discoveries. This is an inevitable caveat when studying a poorly dated period. The definition of the Intermediate category is especially contentious, and it should be stressed that the categories are not intended to be mutually exclusive: Intermediate material could well be contemporary with later-Early or Mid Saxon material, especially outside East Anglia where Ipswich Ware is lacking as an indicator of eighth- to ninth-century dates.

The charred samples are assigned to chronological categories in Appendix 4, following the criteria and terminology set out above. Most assemblages and samples are classifiable as Early, Intermediate or Mid Saxon, with only a small minority being classed as Generic (Table 2).

Table 2 - Chronological distribution of assemblages and samples in the project dataset.

phase	no. assemblages	no. samples
Early Saxon	37	207
Intermediate	22	151
Mid Saxon	39	321
Generic	13	57
Total	**111**	**736**

In two cases, a sample's botanical composition appears potentially to conflict with the artefactual date of its parent context. Thus <SMB1> from Brandon and <AL1> from Alchester both derive from contexts considered on artefactual grounds to be Early Saxon. Yet the botanical composition of <SMB1> is thought to be Iron Age in character, that of <AL1> Romano-British. In both cases, the taxon considered most anomalous in an Early Saxon context is spelt wheat, whose glume bases dominate the cereal component of both of these samples (Booth *et al.* 2001: 202–207; Murphy and Fryer in Tester *et al.* 2014:

28). While we lack radiocarbon readings to confirm the date-range of this material, I have provisionally opted to classify these samples with the more conventional Early Saxon samples: partly because the stratigraphic integrity of the Alchester sample appears to be strong, and partly because spelt does occur in other post-Roman samples (see Chapter 5). Also, the argument that spelt-rich samples cannot be post-Roman, because post-Roman samples are not rich in spelt, is a circular one.

At Lake End Road, Dorney, there is a conflict between the radiocarbon date range of the charred plant remains in <LER1> (cal. AD 430–660), and the eighth- to ninth-century date of the stratigraphically associated pottery. If the radiocarbon determination is accepted, then the plant remains must be assumed to represent residual Early Saxon material in a Mid Saxon deposit (Foreman *et al.* 2002: 58–60). The possibility is thus raised that residual Early Saxon material might also be represented in other presumed Mid Saxon contexts at this site. Provisionally, in the absence of additional data, I have accepted the radiocarbon date for <LER1>, but the remaining samples from the site have retained their published eighth- to ninth-century dates.

Geographical patterns

Charred archaeobotanical data are not evenly distributed throughout the study regions: Cambridgeshire is the best represented county, followed by Oxfordshire, Buckinghamshire, Norfolk and Suffolk. By contrast, Berkshire, Gloucestershire, Wiltshire and Essex are poorly represented (Figure 4). This imbalance is to some extent determined by more general patterns in the distribution of excavated Anglo-Saxon settlements, and does not necessarily, in itself, tell us anything about the plant economy of the case study regions. It does perforce mean, however, that the results in this book will elucidate farming better in Oxfordshire, Buckinghamshire, Cambridgeshire, Norfolk and Suffolk than elsewhere.

This inter-county variation naturally translates to differential representation of the National Character Areas: the Bedfordshire and Cambridgeshire Claylands; Breckland; South Norfolk and High Suffolk Claylands; South Suffolk and North Essex Claylands; Fens; and Upper Thames Clay Vales are all particularly well represented in terms of both assemblages and samples in the project dataset (Table 3).

The geographical distribution of samples is skewed somewhat by four unusually well-represented assemblages, all in Suffolk: Flixton with 30 samples, Eye with 59 samples, Brandon with 50 samples, and Ipswich with 35 samples. The botanical data for Flixton and Eye are, at the time of writing, available only in the form of semi-quantitative assessments, rather than fully quantitative analyses. This fact not only limits the analytical utility of these data, but could also help to explain the relatively high frequency of samples recorded in these assemblages, since assessments are more inclusive than quantified analyses. As advised by Historic England (formerly English Heritage), archaeobotanical assessments are intended to review all samples in an assemblage in order to identify those which may justify fully quantitative analysis. Hence, assessments are inherently broader and less critical in their coverage than are fully quantitative analyses, which exclude samples deemed not to warrant further work: for example, those containing a very low density of plant remains (Campbell *et al.* 2011: 7–8).

Table 3 - Geographical distribution of assemblages and samples in the project dataset, in terms of National Character Areas.

National Character Area	no. assemblages	no. samples
Avon Vale	1	20
Bedfordshire and Cambridgeshire Claylands	19	114
Bedfordshire Greensand Ridge	2	5
Berkshire and Marlborough Downs	1	1
Breckland	12	91
Chilterns	3	9
Cotswolds	2	11
East Anglian Chalk	7	46
Greater Thames Estuary	1	4
Mid Norfolk	1	5
Midvale Ridge	1	4
North East Norfolk and Flegg	1	2
North West Norfolk	2	13
Northern Thames Basin	1	6
Rockingham Forest	1	3
Salisbury Plain and West Wiltshire Downs	3	33
Severn and Avon Vales	1	1
South Norfolk and High Suffolk Claylands	3	90
South Suffolk and North Essex Claylands	8	58
Suffolk Coast and Heaths	1	15
Thames Basin Heaths	1	4
Thames Valley	6	21
The Fens	9	56
Upper Thames Clay Vales	24	124

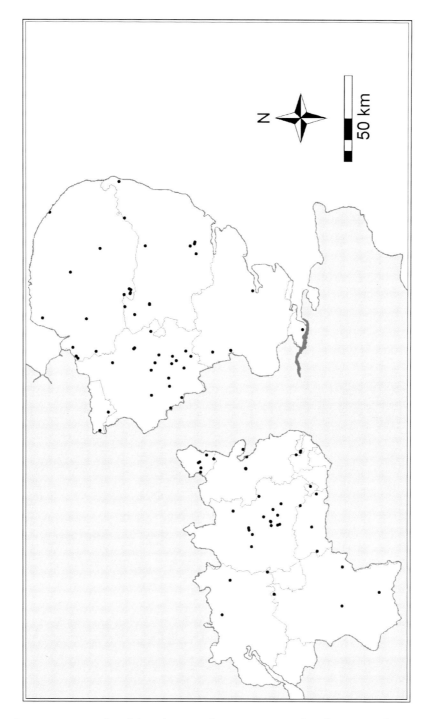

Figure 4 - Geographical distribution of sites represented in the project dataset.

Feature typology

For each sample, a description of its parent context was recorded. Contextual information can assist in the interpretation of samples' depositional histories and botanical contents (Hillman 1981: 124–125). Feature types were therefore recorded where specified in original reports, and then grouped into broad categories to facilitate consistent analysis (Appendix 2: Metadata 1). These standardised categories, although admittedly arbitrary and somewhat subjective in nature, at least provide a broad indication of the types of context from which the samples derive.

Plant parts

Standardised terminology was also applied to a selected suite of plant parts, so that different words for comparable items – e.g. achene, nutlet, pulse – were gathered under a single term (in this case, 'seed') for ease of analysis. A concordance table showing which original terms have been bracketed under a restricted set of standard terms is a key piece of metadata (Appendix 2: Metadata 2). It is important, for repeatability's sake, to detail this information, as not all analysts will necessarily complete this amalgamation exercise in the same way.

Seed was used for all achenes, nutlets and equivalent propagules for quantification purposes. In the case of pulses, where cotyledon counts were given in a report, these were halved (and rounded to the nearest integer) to give the equivalent minimum number of whole seeds thus represented. The standard term rachis was applied to either nodes or internodes, whichever was greater in a given sample, so as to derive the minimum number of rachis segments for a given taxon. The standard term glume base includes not only individually recorded glume bases but also spikelet forks, commuted into two glume bases each for the purposes of comparative quantification. A working assumption was made that all indeterminate rachis segments were likely to represent free-threshing cereals, while indeterminate glume bases were more likely to represent glume wheats, in accordance with predicted preservation biases (Boardman and Jones 1990: 2). Rare instances of rachis ascribed to glume wheats, or glume bases ascribed to free-threshing cereals, were therefore deemed redundant, since the chaff of the respective cereals had already been quantified using the other standard parts described above. Culm nodes constitute another standard category, but only where ascribed to cereals (or large grasses which could represent cereals).

All other recorded plant parts were either very rare, ill-defined, or else of uncertain quantitative value, and have therefore been deemed non-standard parts (e.g. 'chaff', 'flowering stems', and 'seed cases'). Unless otherwise stated, non-standard parts were excluded from quantitative analyses, so as to enhance the comparability of the data. Where relevant, however, the material was considered in semi-quantitative presence analyses of the various taxa.

The quantity of items ascribed to each taxon and plant part was recorded, except in those cases where the available reports stated presence only, or used a semi-quantitative scale of abundance (e.g. 'x' = 1-50, 'xx' = 50-100). These latter samples are unsuitable for quantitative analyses but are nonetheless important in semi-quantitative presence analyses.

Taxonomic nomenclature

Taxonomic identifications were initially recorded verbatim from original reports, including indeterminate items. Standardised nomenclature was then applied across the dataset. For most taxa, nomenclature follows Stace (2010) but, for cereals, terminology was adapted from the older conventions common in archaeobotanical literature, including some 'traditional' names that have since been subject to taxonomic revision (Table 4; Cappers and Neef 2012: 15-16). The traditional names are preferred in this study because of their customary usage in archaeobotanical literature, but the genetic relationships encapsulated in the new scientific names are worthy of note, e.g. spelt and bread wheat belonging to the same biological species and therefore being interfertile. Commentaries on the transition from Romano-British to Anglo-Saxon agriculture, including a transition from spelt wheat to bread wheat cultivation, have generally given little consideration to the possibility of cross-fertilization between spelt and bread wheat. Given that an important debate for this period is why, when and how bread wheat supplanted spelt as the predominant wheat crop in post-Roman Britain, the genetic angle – the possibility of hybridisation – might be worthy of more detailed consideration, but is beyond the scope of this project.

Table 4 – Traditional and new scientific nomenclature of cereals (after Cappers and Neef 2012: 15-16); the traditional names are used in this book.

traditional name	new scientific name	common name
Triticum spelta L.	*Triticum aestivum* ssp. *spelta*	Spelt wheat
Triticum dicoccum Schübl.	*Triticum turgidum* ssp. *diccocon*	Emmer wheat
Triticum aestivum L.	*Triticum aestivum* ssp. *aestivum*	Bread wheat
Triticum turgidum L.	*Triticum turgidum* ssp. *turgidum*	Rivet wheat
Secale cereale L.	*Secale cereale* ssp. *cereale*	Rye
Hordeum vulgare L.	*Hordeum vulgare* ssp. *vulgare*	Six-row barley
Hordeum distichum L.	*Hordeum vulgare* ssp. *distichon*	Two-row barley

'Cereals indet.' was used as a standard term for all indeterminate cereal records, including ambiguous cases such as *Secale/Triticum*. Likewise, 'large legume indet.' was used to denote identifications such as *Vicia/Lathyrus*, *Vicia/Lathyrus/Pisum*, *Vicia/Pisum/Lens*, and similarly ambiguous records. The additional classification '*Triticum* L. free-threshing' was used to embrace the many terms – often synonymous and sometimes misleading – which archaeobotanists have used to describe free-threshing varieties of wheat. This generic term was deemed appropriate for all positive identifications of free-threshing wheats, since individual species such as bread and rivet wheat can be very difficult to distinguish in the archaeobotanical record unless rachis fragments are sufficiently well-preserved (Moffett 1991: 233–235). '*Triticum* L. free-threshing' serves as a more neutral substitute for the widely-used umbrella term *Triticum aestivo-compactum*. Following the widespread practice in much archaeobotanical literature, the term 'glume wheat' is used in this project as a shorthand for emmer, spelt, and ambiguous identifications of 'emmer/spelt' (einkorn is not present). It has here been preferred to the synonymous term 'hulled wheat', so as to avoid potential confusion with hulled barley, since the 'hull' denoted in each case is different.

Having outlined the nature of the project dataset, and the basic protocols used in its compilation, it is time to consider in more detail its taxonomic composition.

Chapter 3: Surveying the Species

This chapter provides an account of the taxa recorded in the project dataset, including cereals – the main focus of this study – plus other field crops, and arable weeds. Observations are made concerning the identification of taxa, their ubiquity or scarcity, and the allocation of them to categories such as crop, weed and non-arable. The main quantitative analyses are reserved for the following chapters.

312 different taxa were initially recorded in the dataset, at varying levels of precision, including families, genera, subgenera, species, subspecies and varieties. Some taxonomic categories incorporate a range of genera or species that a particular record could plausibly represent. Examples include '*Avena/Bromus*' and '*Ranunculus acris/bulbosus/repens*'. Such is the equivocal nature of archaeobotanical material. The result of these ambiguities and varying levels of precision is that some taxonomic categories are likely to overlap – perhaps, in a few cases, entirely. In such cases, categories were amalgamated to avoid taxonomic redundancy. Such amalgamation required at least one of the following conditions to be met:

i. That the seeds of the amalgamated taxa were deemed to be virtually indistinguishable on the basis of gross morphology, e.g. *Bromus hordeaceus* and *Bromus secalinus*.
ii. That the taxonomic identifications were, potentially, entirely overlapping within the project dataset for a given mode of preservation. For example, charred records of *Malva* sp. and *Malva sylvestris* could be amalgamated, since no other species in the genus *Malva* occurred within the charred dataset. Similarly, subspecies and varieties were amalgamated to species level: e.g. *Vicia faba* var. *minor* was amalgamated with *Vicia faba*.

In this way, ecologically comparable taxa were not counted repeatedly, but species-specific information was retained where appropriate. Taxonomic amalgamation is not an objective process that could easily be standardised or parameterised, because it depends on the judgement, experience and research aims of the individual scholar. It is therefore important that a complete inventory of such amalgamations is given alongside the analyses which follow (Appendix 2: Metadata 3).

In this way, the project dataset's taxonomic range was consolidated to 264 different identifications. These are listed in Appendix 5, the inventory of plant taxa, where each is assigned to one of five different groups: (A) likely cultivars, (B) possible cultivars, (C) possible arable weeds, (D) non-arable taxa, and (E) indeterminate. Distinctions between cultivated and weedy taxa are to some extent arbitrary and should be considered provisional. Remains of indeterminate *Avena* L., for example, might equally represent cultivated or wild oats. Conversely, several taxa provisionally classified as possible arable weeds could well represent cultivated or at least deliberately gathered wild resources, such as fat hen (*Chenopodium album* L.). Class D, non-arable plants, consists mostly of woody perennials that are unlikely to set seed under arable conditions. Aquatic taxa were also assigned to class D, but damp-ground plants were categorized as possible arable weeds, to allow for the possibility of their growth in poorly-drained farmland in antiquity (Jones 1988: 89–90).

Semi-quantitative 'presence analyses' were undertaken by calculating the proportion of units – assemblages or samples – within which each taxon occurred. Such presence values offer a broad perspective on vegetation history, useful for characterising general patterns in the relative frequencies of taxa (Hubbard 1980: 52–53). This semi-quantitative level of description does, however, have limitations. Primarily, because it is a binary system (a taxon is either present or absent), it cannot differentiate between major, minor, and negligible occurrences within individual samples or assemblages (Jones 1991: 64–65). A taxon could be ubiquitous without ever being abundant, or abundant in only a small number of samples. Presence analysis will therefore tend to amplify small differences in the archaeobotanical record, and conversely understate large quantitative differences.

In a strict approach, presence analysis should be applied only to samples or assemblages with similar quantities of plant remains, since the likelihood of any taxon occurring becomes greater as the overall abundance of plant remains increases (Hubbard 1980: 52). This would require an arbitrary quorum to be decided at the outset: the omission, for example, of samples with fewer than 30 identified items. In this instance, however, because the presence analyses are being conducted in tandem with a suite of rigorous quantitative approaches in Chapters 4, 5 and 6, I have opted to maximise the scope of the semi-quantitative work by applying no such quorum here. Effectively, this equates to a quorum of one item per sample (Appendix 1: Parameter 2).

The results of a wide-ranging basic presence analysis, conducted on these terms, are provided in Appendix 5. These are the data cited throughout the remainder of this chapter. We turn our attention first to taxon groups A and B – likely and possible cultivars – before considering the division between possible arable weeds (C) and other wild taxa (D).

Cereals

In accordance with predicted biases, cereals constitute the predominant taxonomic grouping in the charred dataset (Green 1982: 43). Cereal remains are present in all charred assemblages, except for the negligible Mid Saxon assemblage from Wicken Bonhunt which contains only a single seed of *Atriplex patula* L. Cereal remains are also present in all but 31 of the charred samples. These 31 samples are all very small, containing fewer than 30 items each, and therefore the absence of cereals from their contents is likely to be due to chance, as the probability of any taxon occurring is reduced as sample size decreases.

The term '*Hordeum* L.' is here used to embrace all varieties of barley recorded in the dataset except for four seeds of tentatively identified wall barley (*Hordeum* cf. *murinum* L.), a wild grass, in <WFR1>. The remaining identifications are assumed, as in their original reports, to represent cultivars: they occur in 92.8% of assemblages and 66% of samples (Appendix 5). The use of the umbrella term '*Hordeum* L.' deliberately omits to distinguish between six-row and two-row, hulled and naked, dense- and lax-eared varieties of cultivated barley. Charred grains and rachis fragments are often insufficiently well preserved to allow such distinctions to be made (Moffett 2011: 351). Positive identifications of naked and two-row varieties of barley are especially rare in the charred dataset. Nine grains of two-row barley (listed as '*Hordeum distichon*') were identified in <WKB1> from Wicken Bonhunt, but with no indication of the criteria by which they were distinguished as such (Jones n.d.: 8, Table 1). In addition, Vaughan-Williams suggests that two-row barley might be present alongside six-row barley in <FOR1> from Forbury House, because of the relatively high ratio of straight to twisted grains in that sample: possible but not definitive evidence of two-row barley, since all of the straight grains could equally represent six-row barley, with twisted grains under-represented or less clearly identifiable in that sample (Vaughan-Williams 2005: 44).

Positive identifications of naked barley grains occur in only four samples in the dataset: <CRM3>, <FOR1>, <MFB14>, and <WFR1>. The remaining records of barley frequently employ generic or tentative nomenclature, such as '*Hordeum* sp.', '*Hordeum sativum*', or '*Hordeum vulgare sensu lato.*' Hence, the evidence available in the dataset tends to support Moffett's view that six-row hulled *Hordeum vulgare* L. was predominant in Anglo-Saxon barley cultivation (Moffett 2011: 351). The more specific identifications of different varieties of barley are too few and too tenuous to admit further analysis. In any case, the teasing apart of different barley species and varieties is not directly relevant to any of the research questions being addressed in this book.

Wheats are variously categorized as indeterminate, free-threshing and glume wheats. Indeterminate wheat is typically the most common type among both assemblages and samples, with free-threshing wheats proving almost as ubiquitous, and glume wheats occurring far less frequently (see Appendix 5). Free-threshing or 'naked' wheats are those whose ears release their grains immediately upon threshing. Their most commonly preserved chaff elements in archaeobotanical deposits are rachis segments, typically counted as nodes or internodes. The grains of glume or 'hulled' wheats, by contrast, remain tightly enclosed in their spikelets even when threshed, and require dehusking before they are released; the most commonly preserved chaff elements of glume wheats in archaeobotanical deposits are glume bases or spikelet forks (Boardman and Jones 1990; Hillman 1981: 131–137, Fig. 4-7).

Two species of glume wheat occur within the charred dataset: emmer (*Triticum dicoccum* Schübl.) and spelt (*T. spelta* L.), besides the generic identification *T. dicoccum/spelta*. It should be noted that, while the glume bases of these two species are often clearly distinguishable, morphological distinctions between their grains may potentially be more ambiguous, and consequently many of the emmer and spelt grains in the dataset are qualified as 'cf.' or 'type' (Hillman *et al.* 1996: 204–206). It is also possible that some grains ascribed to free-threshing wheats on account of their well-rounded shape might in fact be poorly-preserved, short, round spelt grains (Campbell and Straker 2003: 23). However, it is beyond the immediate scope of this project to pursue this possibility further, and positive identifications of free-threshing wheat grains have provisionally been accepted as such unless the analyst has expressed specific doubts.

Genetically, spelt and emmer are distinguished by their ploidy level, an attribute when relates to numbers of chromosome sets and has implications relating to, for example, drought-tolerance and culinary properties (Moffett 1991: 233–234). Spelt is hexaploid, emmer is tetraploid, and there are comparable differences among the free-threshing wheats of medieval England: bread wheat (*Triticum aestivum* L.) is hexaploid, while rivet wheat (*Triticum turgidum* L.) is tetraploid. Among the free-threshing wheat remains in the dataset, however, very few have been positively identified as representing tetraploid varieties: one rachis segment from Bishop's Cleeve is identified as '*Triticum turgidum* type', and a single rachis fragment from Eynesbury is ascribed to tetraploid free-threshing wheat. The tetraploid wheat rachis nodes identified at Lake End Road were all thought to represent emmer: none was thought to represent a free-threshing variety. These rare instances can be taken as a reminder that, despite the generic terminology employed in many archaeobotanical reports (e.g. '*Triticum aestivum sensu lato*') it is not necessary to presume that free-threshing wheats were exclusively hexaploid in this period. In any case, the ambiguity of free-threshing wheat ploidy levels has little impact on the research questions being addressed here. The umbrella category of 'free-threshing wheat' (*Triticum* L. free-threshing) is therefore sufficient for present purposes.

There is a strong possibility that many, if not most, of the indeterminate *Triticum* L. remains do, in fact, represent free-threshing wheats. Not only are the latter far more common than glume wheats among the positively-identified wheats, but also free-threshing varieties are more prone to distortion during the charring process and therefore, arguably, more likely to be rendered indeterminate (Boardman and Jones 1990: 4–5).

Oats (*Avena* L.) were certainly cultivated later in the medieval period, and are considered here to be a potential crop, although in many cases the wild or cultivated status of *Avena* cannot be determined archaeobotanically with any certainty. Indeterminate remains of oat occur throughout the regions and periods represented in the project dataset (64% of assemblages, 34.6% of samples: see Appendix 5). However, species-level distinctions between wild (*A. fatua*, *A. sterilis*) and cultivated (*A. sativa*, *A. strigosa*) varieties are rare and not always reliable, unless based upon well-preserved floret bases (Jacomet 2006: 52). Floret bases can be identified as being of 'cultivated type', i.e. consistent with – but not definitive proof of – the presence of *Avena sativa* L. However, oat grains *per se* are not considered diagnostic. It is unclear as to how Jones distinguishes between the cultivated '*A. sativa*' and the wild '*A. cf. ludoviciana*' (*A. sterilis* L.) at Wicken Bonhunt and Witton (A. Jones n.d.; A. Jones in Lawson 1983: 67–68).

The remaining cultivar/wild oat distinctions recorded in the dataset concern floret bases and are therefore more reliable, but are of very limited distribution. Oat floret bases of 'wild' type are identified in six samples, representing five sites (Bloodmoor Hill, Chadwell St Mary, Ipswich, Rosemary Lane, and Brandon); those of 'cultivated' type occur in only three samples at two sites (Lake End Road and Ipswich). These few records are clearly insufficient to support wider generalizations about the cultivation of oats in the study regions. Since there is no clear way of resolving the ambiguity that surrounds the majority of *Avena* grains recorded in the project dataset, in the following analyses both indeterminate and 'cultivated type' *Avena* records have been amalgamated into a single category and treated as a potential crop, in order to maximize the available crop data with minimal loss of usable taxonomic information.

Grains and rachis segments of rye (*Secale cereale* L.) constitute the least widespread category of free-threshing cereal remains in the dataset (48.6% of assemblages, 26.8% of samples: see Appendix 5). Although occasionally qualified as 'cf.' or simply termed '*Secale*', rye is taken here as a consistent and homogeneous taxonomic category – *Secale cereale* L. – which is the only cultivar species in that genus (Zohary *et al.* 2012: 59–66).

Pulses

Pulses, the edible propagules of large-seeded leguminous plants, cultivated or otherwise, are far rarer than cereals. They occur in only 69.4% of assemblages and 36.5% of samples (Appendix 5). The vast majority of these occurrences are classified as indeterminate large legumes, most often designated '*Vicia/Lathyrus*' in original reports. These records, along with those identified generically as *Vicia* or *Lathyrus*, could represent cultivars (such as *Vicia faba* L.) or wild species (such as *Vicia tetrasperma* (L.) Schreb.). In comparison, positively-identified leguminous cultivars have a negligible presence in the dataset. Lentil (*Lens culinaris* Medik.) is restricted to one seed, qualified 'cf.', in <Y49>, and two in <FOR1>. Moffett raises the possibility that this species could have been grown as a fodder crop, as documented for the post-medieval period (Moffett 2011: 352). The more common garden pea (*Pisum sativum* L.) and broad bean (*Vicia faba* L.) are far from being ubiquitous, and are only ever identified in relatively small quantities. There are rarely more than two seeds of these species in any given sample; the maximum recorded is in <WFR5>, which contains 25 *Vicia faba* seeds.

As will be demonstrated in Chapter 4, no sample has a crop component dominated by cultivated legumes. Given their scarcity, these charred pulses are likely to represent little more than chance contaminants in cereal-dominated assemblages, whether as grain-mimics accidentally harvested with a cereal crop, or as part of mixed processing or cooking waste. They cannot necessarily be considered in any way representative of pulse cropping in Anglo-Saxon England. Given the scarcity of specifically identified leguminous cultivars, the seeds of generic *Vicia*, *Lathyrus* and comparably indeterminate large legumes are treated hereafter as possible weed seeds, rather than as crops.

Flax

Flax (*Linum usitatissimum* L.), which could have been harvested as a fibre crop and/or for the culinary use of its oily seeds, occurs in 15.3% of assemblages and 4.2% of samples (Appendix 5). Particular concentrations are present in the Mid Saxon assemblages from Yarnton and Brandon, occurring in six and nine samples respectively.

There is little ambiguity surrounding the identifications of flax in the dataset. Few are qualified by 'cf.', and alternative *Linum* identifications are comparatively rare: fairy flax (*Linum catharticum* L.), a wild taxon, occurs in only one sample, indeterminate *Linum* L. in just three. The seeds and occasional fragmentary capsules of cultivated flax usually occur only in small quantities, with fewer than 20 items per sample. Flax is well represented in a small number of samples, and even constitutes the dominant crop type in the Mid Saxon sample <SMB16> from Brandon (see below, Chapter 4), but overall the corpus of charred flax macrofossils is too small to be very informative or representative of Anglo-Saxon flax cultivation. It is best used in combination with the somewhat richer waterlogged evidence for flax production and processing, as discussed in the companion volume (McKerracher 2018: 111–113).

Other possible crops

A range of other species could represent cultivars, but register only negligible presence and abundance in the project dataset, and are therefore of very limited informative potential. One such species is hop (*Humulus lupulus* L.), best known as a flavouring agent in brewing, three seeds of which occur in the Mid Saxon sample <IPS11> from Ipswich. Another similar example is hemp (*Cannabis sativa* L.), which could have been utilised for its fibres, seeds, and perhaps psychoactive properties, only three seeds of which are recorded within the dataset: in the Mid Saxon sample <IPS5> from Ipswich. Such evidence is clearly too meagre to support wider speculation on the cultivation and use of hemp and hop in Early and Mid Saxon England.

The same is true of the two charred grape pips (*Vitis vinifera* L.) in the dataset: one in each of the Mid Saxon samples <EN3> from Eynsham and <Y26> from Yarnton. A further pip was noted in the assessment of the Lake End Road assemblage: its parent sample was not fully analysed or published, however, and therefore does not appear in this project's dataset (Pelling in Foreman *et al.* 2002, CD-ROM). Although these three sites are all situated far inland, they were also in receipt of Ipswich Ware and other imported pottery types, an indication of their involvement, via the Thames, in wider trading networks. Hence, the grapes could well have been imported as raisins. The pips alone do not constitute evidence of Mid Saxon viticulture in the study regions.

Opium poppy (*Papaver somniferum* L.) may have been cultivated for its culinary or psychoactive applications, but evidence is limited. Charred seeds occur in four samples across three Mid Saxon assemblages: one in each of <HUT6> and <HUT7> from the Hutchison Site; one in <LW7> from Lake End Road; and 180 '*Papaver* cf. *somniferum*' seeds in <Y47> from Yarnton, although these could, as Stevens suggests, represent arable weeds rather than cultivated plants (Stevens in Hey 2004: 363).

From this limited evidence, there is little that can be inferred about the status of any of the above species as Anglo-Saxon crops in the study regions. Nonetheless, it may be significant that their occurrences are exclusively Mid Saxon in date. They could potentially, therefore, represent crop-biological innovations of the period. In any case, as will become clear in Chapter 4, like flax and pulses these species occur too rarely in the project dataset to admit further quantitative or semi-quantitative analysis. In the event, among the various potential crop species recorded in the dataset, it is only the cereal remains which are sufficiently widespread and abundant to admit extensive, comparative, numerical analysis.

Wild taxa

It is of central importance to this study to define a group of taxa as potential segetal weeds, i.e. wild plants that grew among the crops in arable fields. This is not necessarily a straightforward task, since not all wild plants can be deemed potential arable weeds. While some of those species represented in the dataset are there because their seeds were accidentally harvested along with the crop, and stored and processed and ultimately charred along with that crop, others might have entered the archaeobotanical record by other means. For instance, a wild plant could have been growing as a ruderal weed at a settlement where grain was processed, such that its seeds were part of the general detritus present around homes, hearths, stores, and the like. It may also be that a wild plant was deliberately gathered – as a foodstuff, for instance, or for medicinal use – and arrived at a settlement in that way.

As will be seen in Chapter 4, there are statistical methods for ascertaining whether the wild species represented in an archaeobotanical sample represent a plausible, coherent arable weed flora. For the purposes of this section, it is sufficient to determine which taxa can, on ecological or cultural grounds, be considered potential segetal weeds or a different category of wild plant.

A number of the species classified in this project as possible arable weeds might also represent if not cultivars then at least wild species deliberately gathered for culinary or other uses. For instance, fat hen (*Chenopodium album* L.) is often mooted as a food/fodder plant in antiquity (Stokes and Rowley-Conwy 2002). Fat hen occurs in 38.7% of assemblages and 15.6% of samples (Appendix 5), usually with low abundance (<20 seeds per sample) but occasionally occurring in relatively large quantities (>100 seeds). In one instance it dominates a sample: 508 seeds in the Early Saxon sample <BRT2>, from a well at Brandon Road North, Thetford. This is the only recorded instance where seeds of fat hen far outnumber those of other possible weed species, in addition to outnumbering cereal grains. One could conjecture that this sample represents rare evidence of a burnt store of fat hen seeds (whether gathered by humans or rodents). Otherwise, however, compelling evidence for its cultivation or collection is lacking, and in general it may justifiably be treated as an arable weed.

Another exceptional instance of a potentially useful wild species dominating a sample is found at the Hutchison Site, Cambridge, where 700 tentatively identified seeds of black mustard (*Brassica* cf. *nigra* (L.) W.D.J. Koch) constitute 53.1% of the macrofossils in Mid Saxon sample <HUT5>, derived from a well. Most of the other macrofossils in this sample are cereal grains. It is possible that the sample represents mixed burnt food waste, with the black mustard seeds having been gathered for medicinal purposes or for culinary use as a flavouring agent.

A similarly anomalous sample is <HAM18-20> from Harston Mill, probably of Early Saxon date, whose charred seeds are dominated by cleavers (*Galium* cf. *aparine* L.; 68.4% of 136 seeds). Although normally interpreted as an arable weed, cleavers produces edible parts and could also have had medicinal applications (Launert 1981: 176; O'Brien 2016: 100–101). The burred fruits are, however, notoriously adhesive, such that the seeds' entry to the settlement could have been entirely accidental, whether or not it was growing as an arable weed.

An interesting problem is posed by soft brome and rye brome, which are not easily distinguished on the basis of seed morphology and which have therefore been treated here as a single amalgamated category (*Bromus hordeaceus/secalinus* L.). Many indeterminate *Bromus* records are also likely to represent *B. hordeaceus/secalinus*, since alternative species of *Bromus* are lacking in the dataset. In 1990, Banham noted that brome seeds were often found in association with cereal remains in Anglo-Saxon botanical assemblages, and cautiously raised the possibility that brome grasses may have been deliberately exploited, as demonstrated for some periods of prehistory (Banham 1990: 35; Behre 2008: 70).

In the dataset compiled for this project, seeds of *Bromus hordeaceus/secalinus* (including indeterminate *Bromus*) are indeed unusually widespread for a non-cereal taxon, occurring in 40.5% of assemblages and 17.7% of samples (Appendix 5). However, given that *Bromus hordeaceus/secalinus* seeds are grain mimics and therefore likely to remain with harvested crops whether or not they are desired, the argument for deliberate exploitation can be made only tentatively. Sample <FOR1> from Forbury House, Reading, is particularly notable for its 262 indeterminate *Bromus* seeds, constituting 32.1% of the charred macrofossils, a proportion only exceeded by barley which may, conceivably but not demonstrably, include a rare instance of two-row barley, as discussed above. <FOR1> is also notable for its two lentils, another rare taxon which, as noted above, could perhaps have been a fodder crop. It could thus be conjectured that <FOR1> may represent a burnt fodder store, comprising brome and two-row barley with traces of lentil and other pulses. In general, however, it is reasonable to classify the bromes as potential arable weeds.

A number of other wild taxa, meanwhile, have been deemed non-arable plants: primarily woody perennials that are unlikely to set seed on farmland, but also some aquatic species which even a poorly drained field would struggle to sustain. Certain woody perennials may nonetheless have been exploited for food, whether through arboriculture or gathering from the wild, leading to their presence among charred crop remains. Of these species, the most common is hazel (*Corylus avellana* L.), the fragmented nutshells of which occur in 38.7% of assemblages and 17.8% of samples (Appendix 5). Other, rarer examples include fruiting brambles (*Rubus fruticosus* L. agg.), raspberry (*Rubus idaeus* L.), varieties of cherry, plum and bullace (*Prunus* L.), and elder (*Sambucus nigra* L.). The common club rush (*Schoenoplectus lacustris* (L.) Palla) is a rare instance of an aquatic plant preserved amongst charred plant remains.

Finally, very many samples include a record for 'indeterminate seeds', i.e. those too poorly preserved for the archaeobotanist to identify accurately. These have been grouped separately here, and largely omitted from the quantitative analyses, since they can provide little useful information, other than contributing to the overall quantity of plant remains in a deposit.

This chapter has demonstrated that, out of a diverse range of plants represented in the dataset, cereal crops are ubiquitous, and most other taxa can be deemed potential arable weeds. The project dataset is thus ideally suited for the investigation of cereal cultivation in Early and Mid Saxon England.

Chapter 4: Defining the Deposits

As discussed in Chapter 1, a core element of Mid Saxon agricultural development is thought to have been an increase in surplus crop production, particularly with regard to cereals. According to this model, levels of grain production and processing were considerably higher by the 8th and 9th centuries than they had been in the 5th and 6th. Various kinds of evidence – watermills, pollen cores, sedimentation sequences – can be cited in support of this model, but none is particularly widespread within the case study regions. The evidence of charred plant remains is ubiquitous by comparison. What can these remains tell us about scales of production and processing in the Early and Mid Saxon periods?

It might seem much too simplistic to suggest that larger amounts of charred grain can be indicative of increased surpluses. Surely the taphonomic gulf between the farmer's field and the archaeobotanist's petri dish is too great to admit a simple correlation of scale? At the level of individual samples, this scepticism is well deserved. For a large group of samples considered collectively, however, there are in fact good reasons to posit some positive correlation between surplus production and archaeobotanical abundance. This is because charred plant remains are a routine (though accidental) by-product of cereal processing (van der Veen 2007). From the perspective adopted here, it is not unreasonable to infer an increase in cereal processing activity from the increased occurrence of charred plant remains at settlements over a given period. More specifically, in the words of van der Veen and Jones (2006: 222): 'The answer to the question, "where are accidents involving parching, drying and storage most likely to occur?" is that they will tend to occur in places where these activities are regularly carried out, i.e. where grain is handled in bulk.'

Even if we accept this premise, however, there remains the question of how to measure the occurrence of charred plant remains as a proxy for levels of bulk grain handling. For van der Veen and Jones, the key index is the presence of large grain-rich samples. While charred plant remains *per se* are routinely made at cereal-handling settlements, large grain-rich deposits – representing valuable, part-processed food stores – are only likely to have been created on any appreciable scale in the context of especially frequent and large-scale bulk handling (van der Veen and Jones 2006: 226).

Nonetheless, the three variables of frequency, largeness and grain-richness are difficult to define in absolute terms. There is no ideal definition of a 'large grain-rich sample' which can be sought in the archaeobotanical records of different places and periods. Rather, to account for the particular characteristics of Early and Mid Saxon archaeobotany, we must look for synchronic and diachronic trends within the confines of the project dataset. To begin with, following van der Veen and Jones' lead, we must interpret the samples as artefacts of crop processing, through the application of quantitative analyses. This is not only a valuable exercise to prime the dataset for the investigation of surplus production; it also acts as a preliminary data-cleaning step for the more taxonomically detailed analyses of crop and weed species which follow in Chapters 5 and 6.

The value of quantification

Hubbard rejected fully-quantitative descriptions of botanical data in favour of semi-quantitative approaches – i.e. presence analysis – on the grounds that the numerical composition of an archaeobotanical deposit is unlikely to be an accurate reflection of its 'parent economy', given the wide range of taphonomic biases to which it has been subject, in terms of deposition, preservation and recovery (Hubbard 1980: 51). In some cases, where the impact of these potential biases is impossible to determine, presence analysis may indeed be the most legitimate use of the data. However, measures can often be taken to identify and account for some of these taphonomic factors. Certain preservation biases, such as the differential effects of charring on different taxa and plant parts, are predictable and can to some extent be accounted for simply by standardising the units of analysis and by isolating different modes of preservation (Jones 1991: 64, 69). This has already been achieved by restricting the project dataset to charred plant remains in plausibly independent samples (see Chapter 2).

Another potentially biasing factor, crop processing, has been found to influence not only the crop content of samples, but also the weed floras preserved alongside the crops. In short, the sequential stages, whereby a harvested crop is processed to create a usable commodity, introduce systematic biases into the botanical composition of samples, biases which might produce spurious ecological or economic trends in the resultant data (Bogaard *et al.* 2005: 508). However, these biases are also predictable and the problem can be addressed mathematically. The objective here is to classify samples in terms of which crop processing stage or stages, if any, they are likely to represent. Strictly speaking, only those samples representing the same basic stages of the crop processing sequence should be included in comparative analyses, since they are, in archaeobotanical terms, the same kinds of artefact.

Dominant crop types

The principal crop types in the dataset are processed in different ways: cereals, flax and pulses each require a distinct sequence of processing stages to render them economically useful. Among cereals, too, there are variant processing sequences for glume wheats and free-threshing cereals. Samples were therefore first grouped according to their dominant crop types. A crop type was considered dominant if its standard parts constituted at least 80% of all positively-identified standard crop items in the sample (Appendix 1: Parameter 3), provided that the standard crop items numbered at least 30 in total (Appendix 1: Parameter 4). Standard items are those listed as such in Appendix 2 (Metadata 2).

The analysis was conducted in two stages. First, for those samples with at least 30 seeds/grains, rachis segments, and glume bases belonging to cereals, pulses, flax and the 'other crops' described in Chapter 2, I calculated the relative proportions of those items for each of those crop categories. 198 samples were included in this step, of which 194 were dominated by cereals, one was dominated by flax, and three were not clearly dominated by any single category but rather by a mixture of cereal and flax. Clearly, then, the only samples worthy of further comparative analysis are those 194 dominated by cereals, there being too few of any others to admit meaningful comparisons.

The second stage of this analysis took those 194 samples whose dominant crop type was cereals, and which contained at least 30 rachis segments, glume bases, and grains which could be assigned either to a free-threshing cereal or to glume wheats. Grains identified only as 'cereals indet' or 'wheat indet' could not be assigned to either of these cereal types and were therefore omitted from this stage of analysis. However, following the logic outlined in Chapter 2, indeterminate rachis segments were taken to represent free-threshing cereals, and indeterminate glume bases were taken to represent glume wheats. Of the 136 samples thus included, 118 were dominated by the grains and rachis segments of free-threshing cereals; three were dominated by the grains and glume bases of glume wheats; and 15 were not clearly dominated by either category of cereals, although free-threshing cereals are the best represented category in most of these 'mixed cereal' samples.

Hence, by a significant margin, the 118 samples in which free-threshing cereals are the dominant crop type are those best suited to further comparative analysis. These will henceforth be termed the free-threshing samples, for brevity.

Investigating crop processing

Having isolated a large group of samples dominated by a common crop type, we may now proceed to the analysis of crop processing stages. As discussed above, the processing of harvested crops introduces a taphonomic bias into the botanical composition of samples, so in order to make meaningful comparisons between samples it is important first to determine what sort of behavioural activity – what crop processing stage, if any – is represented by each sample. A sample's constituent crop components and weed seeds may all be affected, not only by the various processing stages themselves (e.g. winnowing), but also by the possible admixture of different products and by-products, which would produce problematic composite samples with potentially misleading co-occurrences of taxa. At the same time, this kind of analysis will show which samples represent the winnowed and sieved 'grain-rich' category highlighted by van der Veen and Jones as an index of surplus grain handling (van der Veen and Jones 2006).

The archaeobotanical study of crop processing has been developed extensively by Glynis Jones, based upon her ethnographic studies of traditional farming communities on the Aegean island of Amorgos, and with reference to similar work in Turkey by Gordon Hillman. Her approach is based upon the observation that the basic processing sequence for cereal crops admits little variation, such that modern traditional practices can provide a useful analogue for past activity. The various stages of threshing, winnowing and sieving are likely to be common to cereal farmers widely separated in time and space (Hillman 1981: 126–138; Jones 1984: 46).

As Hillman has argued, among the various products and by-products of the cereal processing sequence, five should in theory be particularly prone to archaeological preservation by charring: (i) the waste by-products of winnowing, (ii) the waste by-products of coarse-sieving, (iii) the waste by-products of fine-sieving, which may be burnt as fuel; (iv) semi-clean stored grain, i.e. that which has not yet been fine-sieved, and (v) fully clean stored grain, which may be charred accidentally during storage or culinary preparation (Hillman 1984: 11).

For free-threshing cereal crops, winnowing and coarse-sieving by-products should generally be relatively rich in rachis segments; fine-sieving by-products should be dominated by arable weed seeds, except for those which mimic crops and are thus retained by the sieves; and products (iv) and (v) should both be dominated, to a greater or lesser extent, by cereal grain, with comparatively small amounts of chaff and weed seeds.

Such general observations have been worked by Jones (1990) into a quantified model, in which the relative proportions of cereal grains, rachis segments and weed seeds are calculated for each sample dominated by free-threshing cereals. In this book, I have required that each sample in this analysis contains at least 30 of these items in total (Appendix 1: Parameter 5).

The resultant ratios are then compared with those calculated for Jones' ethnographic control samples of known origin, representing winnowing by-products, coarse-sieving by-products, fine-sieving by-products, and fine-sieved products (although Jones found that winnowing and coarse-sieving by-products are not easily distinguishable on this basis). I have summarised the numerical criteria for classifying samples on this basis in Table 5, based upon the published method and ethnographic data (Jones 1990). The results can also be assessed visually by plotting the samples on tripolar graphs

(Figures 5 and 6). The classes 'USG' (unsieved grain) and 'MS' (mixed stages) are not defined by Jones, but represent my own interpolations from her model.

Table 5 – Characterisation of crop processing products and by-products according to constituent proportions of grain, chaff and weed seed (after Jones 1990: 93-96).

product/by-product	grain	rachis	weed seed
FSP (fine-sieved products)	≥80%	≤5%	
USG (unsieved grain, i.e. FSP+FSBP prior to fine-sieveing)	<80%	≤5%	15-50%
FSBP (fine-sieving by-products)		≤5%	≥50%
CWBP (coarse-sieving or winnowing by-products)		>30%	
MS (mixed stages)		6-30%	

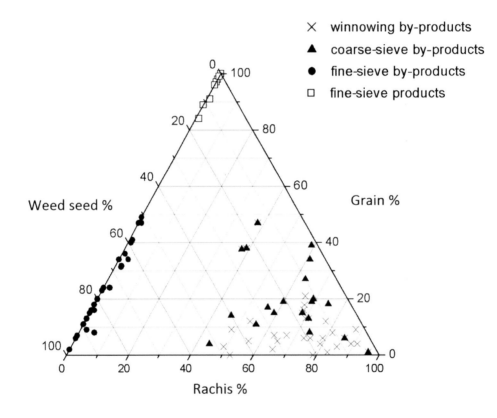

Figure 5 – Tripolar graph of grain : rachis : weed seed ratios in ethnographic control samples analysed by Jones (1990: 93-96).

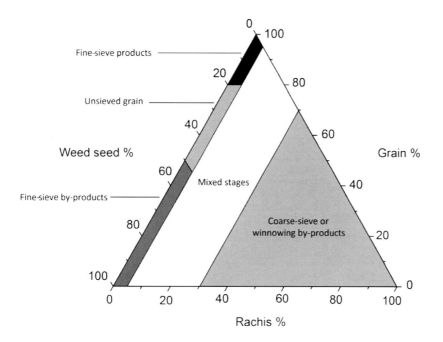

Figure 6 – Idealized interpretation of grain : rachis : weed seed ratios in terms of crop processing products and by-products (after Jones 1990).

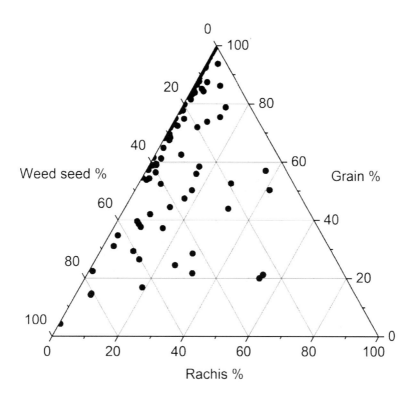

Figure 7 – Tripolar graph of grain : rachis : weed seed ratios of free-threshing cereal samples in the project dataset.

Table 6 - Crop processing analysis of free-threshing cereal samples by basic composition ratios.

sample	interpretation	% grain	% rachis	% weed seed	total items
BCL3	CWBP	43.9	31.7	24.4	542
BEL1-12	FSP	100.0	0.0	0.0	218
BEL14-19	FSP	97.2	0.0	2.8	216
BMH2	FSP	87.2	0.0	12.8	78
CHM1-3	MS	24.5	25.1	50.4	383
CHO1	FSP	84.3	3.9	11.8	51
ENB1	MS	71.9	8.2	19.9	146
ENB3	USG	68.3	1.7	29.9	344
FOR1	USG	57.2	0.6	42.2	797
GAM2	USG	72.4	1.9	25.7	482
GAM4	MS	52.4	6.8	40.8	103
GAM5	MS	52.6	28.2	19.2	546
HAM10	FSP	99.0	0.0	1.0	104
HAM11	USG	77.3	0.0	22.7	132
HAM12	FSP	97.6	0.0	2.4	1085
HAM15	FSP	98.6	0.0	1.4	289
HAM27	FSP	90.0	0.0	10.0	970
HAM28	FSP	98.7	0.0	1.3	303
HAM33	FSP	95.9	0.0	4.1	1157
HAM5	USG	58.3	0.0	41.7	192
HAM8	FSP	98.7	0.0	1.3	77
HAM9	FSP	91.2	0.0	8.8	57
HIL1	MS	75.4	13.5	11.1	126
HUT3	USG	73.3	0.3	26.3	1181
HUT5	FSBP	31.0	3.2	65.7	1269
HUT6	USG	61.2	2.5	36.3	1726
HUT7	MS	58.4	15.6	26.0	308
HUT9	MS	73.7	10.3	15.9	232
IPS1	CWBP	20.0	53.3	26.7	45
IPS11	USG	59.3	1.9	38.8	322
IPS12	USG	54.6	0.0	45.4	130
IPS19	FSP	83.4	0.9	15.7	337
IPS2	FSP	83.9	0.5	15.6	411
IPS20	FSP	93.1	0.0	6.9	130
IPS21	FSP	94.7	0.0	5.3	416
IPS22	FSP	88.6	0.0	11.4	149
IPS23	FSP	93.8	3.6	2.7	112
IPS25	FSP	90.7	0.0	9.3	75

IPS26	FSP	85.0	0.0	15.0	133
IPS28	FSP	87.9	0.0	12.1	107
IPS29	FSP	91.7	0.0	8.3	72
IPS3	USG	71.7	0.0	28.3	515
IPS30	FSP	85.8	0.0	14.2	120
IPS32	USG	68.2	0.0	31.8	85
IPS34	USG	58.8	0.9	40.4	342
IPS38	FSP	94.7	0.0	5.3	113
IPS39	FSP	89.2	0.0	10.8	111
IPS4	FSP	80.0	0.0	20.0	80
IPS5	FSP	83.3	0.0	16.7	227
IPS6	FSP	83.2	0.0	16.8	143
IPS7	FSBP	14.3	4.6	81.1	2165
IPS9	FSP	83.2	0.9	15.9	435
LBQ1-3	FSP	92.9	0.4	6.8	561
LE1	USG	74.9	2.7	22.5	187
LE3	MS	29.3	10.0	60.7	580
LE5	MS	37.6	8.1	54.3	210
LE6	FSBP	14.7	4.6	80.8	1091
LE7	FSBP	4.3	0.4	95.3	1399
LH1	MS	78.8	13.5	7.7	888
LH2	MS	55.9	15.9	28.2	195
LH4	USG	56.4	3.2	40.4	94
LLC10	FSP	83.8	1.5	14.7	136
LLC11	MS	62.5	8.0	29.5	112
LLC22	USG	54.3	2.3	43.4	129
LLC3	FSP	87.6	0.0	12.4	113
LW1	MS	47.4	16.5	36.1	310
LW5	CWBP	50.3	41.2	8.5	636
LW6	FSBP	34.7	2.6	62.7	386
ML4	MS	86.2	8.1	5.7	123
RFT6-7	FSP	84.1	0.0	15.9	214
RFT8-9	FSP	89.8	0.0	10.2	127
ROS1	MS	39.6	6.0	54.4	467
ROS2	FSP	88.7	0.4	10.9	497
ROS4	FSP	92.3	0.5	7.1	182
SMB12	CWBP	57.0	36.7	6.3	158
SMB13	USG	75.0	0.0	25.0	775
SMB17	FSP	98.3	0.0	1.7	116
SMB3	USG	69.6	0.9	29.6	115

TP1	MS	38.7	6.9	54.4	434
TSC2	MS	28.5	28.5	43.1	130
TSC4	CWBP	21.7	31.8	46.5	157
TSC7	MS	16.8	18.9	64.2	95
WAL1	FSP	100.0	0.0	0.0	236
WFC1	FSP	87.7	0.9	11.4	114
WFC11	FSP	90.5	0.0	9.5	95
WFC14	FSP	88.4	0.0	11.6	95
WFC6	FSP	87.4	3.5	9.1	373
WFR1	USG	67.5	1.9	30.7	424
WFR4	FSP	86.6	0.7	12.7	134
WFR5	USG	58.9	2.1	38.9	285
WKB1	FSP	93.4	0.0	6.6	226
WLP10	CWBP	21.2	53.8	25.0	104
WLP2	MS	37.2	15.0	47.8	113
WLP3	MS	42.0	8.7	49.3	69
WLP7	USG	53.8	1.7	44.4	234
WLP8	MS	44.4	13.6	42.0	81
WLP9	MS	50.2	17.4	32.4	259
WRS1	FSP	97.6	0.0	2.4	126
WRS4	FSP	97.4	0.0	2.6	740
WRS5	FSP	82.6	0.0	17.4	207
WRS7	FSP	94.3	0.0	5.7	87
WRS8	FSP	99.2	0.0	0.8	125
WST2	USG	61.5	0.0	38.5	496
WTN1	FSP	99.0	0.0	1.0	103
WTN2	FSP	98.3	0.0	1.7	360
WWI7	MS	26.4	13.2	60.3	242
Y12	FSP	81.4	1.4	17.1	70
Y22	USG	64.8	1.4	33.8	71
Y26	USG	72.3	0.0	27.7	119
Y30	USG	77.7	0.9	21.4	930
Y31	USG	73.7	0.0	26.3	399
Y34-5	USG	79.8	0.3	19.9	719
Y38	FSP	84.9	0.6	14.5	179
Y39	FSP	82.4	0.3	17.4	697
Y44	FSP	85.1	3.0	11.9	67
Y47	FSBP	22.4	1.1	76.5	19829
Y49	FSP	84.4	0.0	15.6	295
Y5	FSP	84.4	0.0	15.6	64

All 118 of the free-threshing samples are eligible for this basic compositional analysis, each having at least 30 grains, rachis segments and weed seeds in total. Of these, 57 are thus classified as fine-sieved products; 26 as unsieved grain; six as fine-sieving by-products; six as coarse-sieving or winnowing by-products; and 23 as mixed stages (Table 6; Figure 7).

As Jones notes, however, a close match between ethnographic and archaeological data cannot necessarily be expected. For example, rachis segments may be under-represented in charred samples because of their relative fragility, a bias demonstrated through experimental charring (Boardman and Jones 1990). Other variables, such as the thoroughness of winnowing and sieving, and the amount of weed material originally harvested with the crop, always remain unknown in archaeological samples (Jones 1990: 92–93).

Given these caveats, it is prudent to complement the basic compositional approach with an independent method of crop processing analysis, again devised by Jones (1987) through ethnographic studies on Amorgos: discriminant analysis of weed seed types. At each processing stage, the removal or retention of weed seeds among the harvested grain depends upon their physical and aerodynamic properties, defined by Jones in terms of size, weight and headedness. Combinations of these three characteristics result in Jones' six seed-types: BHH (big, headed, heavy), BFH (big, free, heavy), SHH (small, headed, heavy), SHL (small, headed, light), SFH (small, free, heavy), and SFL (small, free, light).

Using Jones' original assignments of species to seed-types, along with those used in later studies as well as my own observations, I have assigned weed taxa in the project dataset to the six different seed categories (Jones 1987: 313, Table 1; van der Veen 1992: 207, Table 7.4; A. Bogaard and M. Charles pers. comm.). Not all weed taxa in the dataset could be classified in this way, either because the taxonomic identification was not specific enough (e.g. 'Poaceae'), or because I could not determine an appropriate classification for a particular seed. The classifications that I have used are presented in Appendix 2: Metadata 4. The provision of these metadata is essential for repeatability, since other analysts may use different classifications depending upon their own judgements.

These weed seed types can be used as variables to distinguish between different products and by-products. By using the multivariate technique of discriminant analysis to derive two new variables (discriminant functions) that most strongly differentiate between products and by-products of known origin from Amorgos, Jones demonstrated the possibility of classifying the archaeological samples alongside the modern control samples on the basis of their weed seed contents (Jones 1984: 54–59). Discriminant analysis of the samples in this study was conducted using IBM SPSS Statistics version 25, applying the 'leave-one-out' option for greater rigour (IBM Corporation 2017; ethnographic data used by kind permission of G. Jones). This discriminant analysis utilised all samples in which at least ten weed seeds could be assigned to one of Jones' six types (Appendix 1: Parameter 6).

The input data for each sample consisted of six fields, one for each of the six seed-types. The percentage of classifiable seeds belonging to each seed-type was calculated for each sample. These values were then square-rooted to give the discriminating variable found by Jones to be the most effective (Jones 1984: 49).

The reclassification of control samples achieved a high degree of accuracy (84%). The software provided classifications for the archaeological samples in tabular form. The results were also assessed visually as a scattergraph using the two discriminant functions returned for each sample: Function 1 on the x-axis, Function 2 on the y-axis. Figure 8 shows an interpretative model for this kind of graph. In this model, each circle encloses approximately 90% of the ethnographic control samples representing a particular product or by-product from Jones' original study, and is centred on the 'group centroid' returned by the discriminant analysis for that product or by-product (Jones 1987: 315, Fig. 1). Archaeological samples that fall within or close to these circles may be interpreted accordingly. The circle for unsieved grain – an interpolated category used in this book – is hypothetically positioned between the fine-sieve products and by-products, but it should be stressed that this circle is not derived from Jones' study and therefore not based directly upon her ethnographic control data.

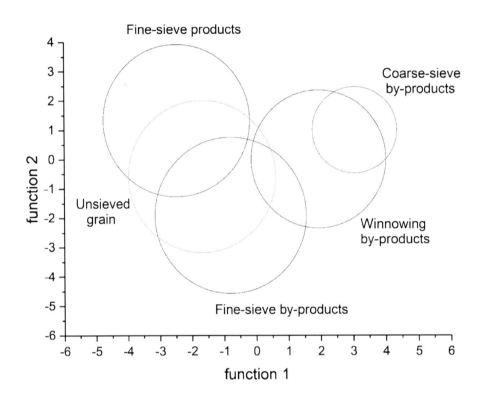

Figure 8 – Model scattergraph for the visual interpretation of discriminant analysis results, after Jones 1987: 315, Figure 1.

Of the 118 free-threshing samples, 89 were eligible for the discriminant analysis of weed seed types. 37 of these were classified as fine-sieved products, 44 were classified as fine-sieving by-products, and eight as winnowing by-products (Table 7; Figure 9).

Table 7 – Crop processing analysis of free-threshing cereal samples by discriminant analysis of weed seed types (discriminant functions rounded to three decimal places).

sample	interpretation	Function 1	Function 2
BCL3	FSP	-1.144	1.093
CHM1-3	FSBP	-1.927	-3.727
ENB1	FSP	-1.334	0.119
ENB3	FSP	-2.475	-0.748
FOR1	FSP	-2.530	0.749
GAM2	FSBP	-1.399	-0.511
GAM4	FSP	-3.249	-1.116
GAM5	FSBP	-2.322	-1.520
HAM11	FSP	-1.897	0.805
HAM12	FSP	-2.449	-0.722
HAM27	FSBP	-3.183	-1.585
HAM33	FSP	-2.007	0.373
HAM5	FSP	-2.143	0.888
HIL1	FSBP	-0.622	-2.755
HUT3	FSBP	-3.395	-2.347
HUT5	FSBP	-1.803	-3.084
HUT6	FSBP	-3.847	-3.404
HUT7	WBP	-0.556	-0.709
HUT9	FSP	-2.417	0.427
IPS11	FSP	-2.976	0.323
IPS12	FSBP	0.066	-2.187
IPS19	FSP	-4.120	-1.226
IPS2	FSP	-4.052	1.314
IPS21	FSP	-2.125	1.270
IPS22	FSP	-2.219	1.257
IPS26	FSP	-1.004	0.445
IPS28	FSBP	-2.177	-1.154
IPS3	FSBP	-3.044	-1.348
IPS30	FSP	-0.550	1.631
IPS32	FSP	-2.639	-0.713
IPS34	FSBP	-1.742	-2.454
IPS39	FSP	-2.815	2.912
IPS4	FSP	-2.371	0.213
IPS5	FSP	-3.019	-0.257
IPS6	FSP	-2.208	-0.125
IPS7	FSBP	-2.661	-3.113
IPS9	FSP	-3.467	-0.093
LBQ1-3	FSBP	-1.096	-3.155

LE1	FSP	-2.508	1.195
LE3	WBP	-1.250	-1.320
LE5	FSBP	-1.733	-1.090
LE6	WBP	-0.026	-1.875
LE7	WBP	0.633	-1.322
LH1	FSP	-3.717	-1.410
LH2	FSBP	-2.317	-0.995
LH4	FSP	-2.426	-0.398
LLC10	FSP	-1.839	0.531
LLC11	FSBP	-1.543	-0.820
LLC22	FSP	-1.902	0.735
LLC3	FSBP	-1.171	-2.620
LW1	FSBP	-3.161	-1.459
LW5	FSBP	-2.319	-1.444
LW6	FSBP	-1.853	-2.423
RFT6-7	FSBP	-2.663	-0.875
ROS1	FSBP	-1.305	-2.298
ROS2	FSBP	-1.466	-0.531
SMB13	FSBP	-1.776	-4.204
SMB3	FSP	-2.744	-0.567
TP1	FSBP	-2.566	-1.525
TSC2	FSBP	-1.303	-4.586
TSC4	FSBP	-1.901	-4.409
TSC7	FSBP	-1.541	-5.044
WFC1	FSP	-1.435	0.061
WFC14	FSP	-0.437	2.575
WFC6	FSBP	-1.628	-0.615
WFR1	FSP	-1.576	1.557
WFR4	FSBP	-2.302	-1.594
WFR5	WBP	-1.033	-0.077
WKB1	FSP	-0.843	0.936
WLP10	FSBP	-1.761	-3.179
WLP2	FSBP	-2.246	-5.165
WLP3	FSBP	-1.896	-5.130
WLP7	FSBP	-2.444	-5.849
WLP8	FSBP	-2.409	-4.342
WLP9	FSBP	-2.294	-5.282
WRS4	FSP	-1.259	0.145
WRS5	FSBP	-1.509	-3.370
WST2	FSP	-1.810	-0.383
WWI7	FSBP	-2.094	-5.086

Y12	FSP	-1.375	0.110
Y22	FSBP	-1.461	-2.106
Y26	FSBP	-0.623	-0.559
Y30	FSBP	-2.051	-1.023
Y31	WBP	0.384	-0.263
Y34-5	FSBP	-0.958	-1.005
Y38	WBP	-0.081	0.592
Y39	FSBP	-1.756	-2.042
Y47	WBP	-0.061	-0.158
Y49	FSP	-2.727	0.564

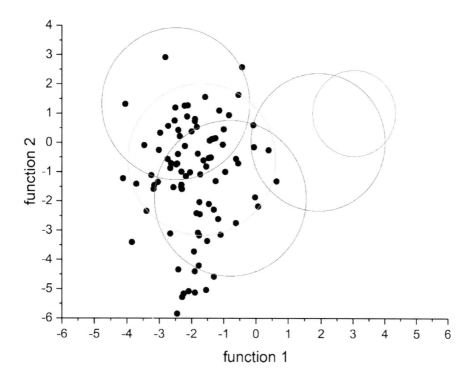

Figure 9 – Discriminant analysis scattergraph of free-threshing cereal samples.

By combining the results of this discriminant analysis with those of the basic ratio analysis described above, we can take account of both the weed seed types and the cereal content of the samples, to derive a more robust crop processing classification. Where the results of the two analyses were compatible for a given sample, that sample was assigned to a crop processing stage accordingly. The criteria for compatibility are summarised in Table 8. Where the results achieved by these two independent methods are mutually consistent for a given sample, it can be argued strongly that the crop and weed contents of that sample 'belong together', i.e. that such a sample could theoretically represent a single product or by-product from the processing sequence, and thus that the taxa therein could have been processed together, and even, theoretically, have grown together (Bogaard 2011: 151).

Table 8 - Theoretical compatibility of ratio and discriminant analysis crop processing classifications.

		Basic composition ratios				
		CWBP	FSBP	USG	FSP	MS
Discriminant analysis	CWBP	X				
	FSBP		X	X		
	FSP			X	X	

Of the 89 free-threshing samples for which both discriminant analysis and basic composition classifications were attainable, compatible results were obtained for 47 (Table 9; Figure 10): three could be classed as fine-sieving by-products (FSBP), 24 as unsieved grain (USG), and 20 as fine-sieved products (FSP). In Figure 10, it is evident that at least three of the supposed USG samples are outliers, falling clearly outside the interpolated boundary circle for such samples. However, given the somewhat hypothetical nature of this interpolated discriminant analysis category, I have opted to persist with the USG classification of these samples for the purposes of this book.

The 44 samples represented by the USG and FSP categories are, by definition, those which could justifiably be deemed to represent 'grain-rich' deposits, and which are therefore of most interest in assessing surplus grain-handling, after the fashion of van der Veen and Jones (2006), as well as constituting a coherent group of artefacts suitable for other comparative analyses. However, for the purposes of tracing surplus crop production and processing, we are required specifically to identify *large* grain-rich samples, and it is to this problem that we now turn.

Table 9 - Combined interpretation of crop processing analyses.

sample	ratio-based interpretation	discriminant analysis interpretation	combined interpretation
BCL3	CWBP	FSP	n/a
BEL1-12	FSP	n/a	n/a
BEL14-19	FSP	n/a	n/a
BMH2	FSP	n/a	n/a
CHM1-3	MS	FSBP	n/a
CHO1	FSP	n/a	n/a
ENB1	MS	FSP	n/a
ENB3	USG	FSP	USG
FOR1	USG	FSP	USG
GAM2	USG	FSBP	USG
GAM4	MS	FSP	n/a
GAM5	MS	FSBP	n/a
HAM10	FSP	n/a	n/a
HAM11	USG	FSP	USG
HAM12	FSP	FSP	FSP
HAM15	FSP	n/a	n/a
HAM27	FSP	FSBP	n/a
HAM28	FSP	n/a	n/a
HAM33	FSP	FSP	FSP
HAM5	USG	FSP	USG
HAM8	FSP	n/a	n/a
HAM9	FSP	n/a	n/a
HIL1	MS	FSBP	n/a
HUT3	USG	FSBP	USG
HUT5	FSBP	FSBP	FSBP
HUT6	USG	FSBP	USG
HUT7	MS	WBP	n/a
HUT9	MS	FSP	n/a
IPS1	CWBP	n/a	n/a
IPS11	USG	FSP	USG
IPS12	USG	FSBP	USG
IPS19	FSP	FSP	FSP
IPS2	FSP	FSP	FSP
IPS20	FSP	n/a	n/a
IPS21	FSP	FSP	FSP
IPS22	FSP	FSP	FSP
IPS23	FSP	n/a	n/a
IPS25	FSP	n/a	n/a
IPS26	FSP	FSP	FSP

IPS28	FSP	FSBP	n/a
IPS29	FSP	n/a	n/a
IPS3	USG	FSBP	USG
IPS30	FSP	FSP	FSP
IPS32	USG	FSP	USG
IPS34	USG	FSBP	USG
IPS38	FSP	n/a	n/a
IPS39	FSP	FSP	FSP
IPS4	FSP	FSP	FSP
IPS5	FSP	FSP	FSP
IPS6	FSP	FSP	FSP
IPS7	FSBP	FSBP	FSBP
IPS9	FSP	FSP	FSP
LBQ1-3	FSP	FSBP	n/a
LE1	USG	FSP	USG
LE3	MS	WBP	n/a
LE5	MS	FSBP	n/a
LE6	FSBP	WBP	n/a
LE7	FSBP	WBP	n/a
LH1	MS	FSP	n/a
LH2	MS	FSBP	n/a
LH4	USG	FSP	USG
LLC10	FSP	FSP	FSP
LLC11	MS	FSBP	n/a
LLC22	USG	FSP	USG
LLC3	FSP	FSBP	n/a
LW1	MS	FSBP	n/a
LW5	CWBP	FSBP	n/a
LW6	FSBP	FSBP	FSBP
ML4	MS	n/a	n/a
RFT6-7	FSP	FSBP	n/a
RFT8-9	FSP	n/a	n/a
ROS1	MS	FSBP	n/a
ROS2	FSP	FSBP	n/a
ROS4	FSP	n/a	n/a
SMB12	CWBP	n/a	n/a
SMB13	USG	FSBP	USG
SMB17	FSP	n/a	n/a
SMB3	USG	FSP	USG
TP1	MS	FSBP	n/a
TSC2	MS	FSBP	n/a

TSC4	CWBP	FSBP	n/a
TSC7	MS	FSBP	n/a
WAL1	FSP	n/a	n/a
WFC1	FSP	FSP	FSP
WFC11	FSP	n/a	n/a
WFC14	FSP	FSP	FSP
WFC6	FSP	FSBP	n/a
WFR1	USG	FSP	USG
WFR4	FSP	FSBP	n/a
WFR5	USG	WBP	n/a
WKB1	FSP	FSP	FSP
WLP10	CWBP	FSBP	n/a
WLP2	MS	FSBP	n/a
WLP3	MS	FSBP	n/a
WLP7	USG	FSBP	USG
WLP8	MS	FSBP	n/a
WLP9	MS	FSBP	n/a
WRS1	FSP	n/a	n/a
WRS4	FSP	FSP	FSP
WRS5	FSP	FSBP	n/a
WRS7	FSP	n/a	n/a
WRS8	FSP	n/a	n/a
WST2	USG	FSP	USG
WTN1	FSP	n/a	n/a
WTN2	FSP	n/a	n/a
WWI7	MS	FSBP	n/a
Y12	FSP	FSP	FSP
Y22	USG	FSBP	USG
Y26	USG	FSBP	USG
Y30	USG	FSBP	USG
Y31	USG	WBP	n/a
Y34-5	USG	FSBP	USG
Y38	FSP	WBP	n/a
Y39	FSP	FSBP	n/a
Y44	FSP	n/a	n/a
Y47	FSBP	WBP	n/a
Y49	FSP	FSP	FSP
Y5	FSP	n/a	n/a

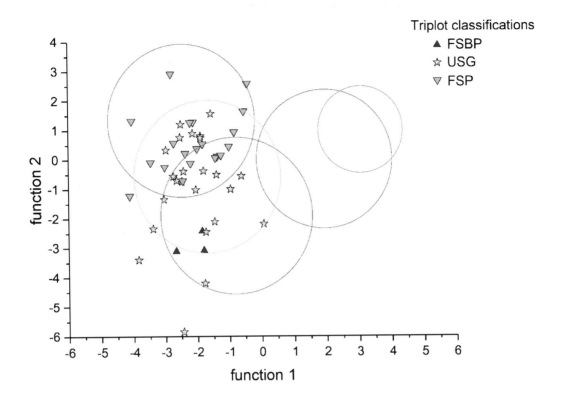

Figure 10 – Discriminant analysis scattergraph of free-threshing cereal samples deemed to have a classification compatible with their basic composition ratios. The symbols used here denote the classifications obtained through the ratio analysis above (Table 6).

Abundance and density

There is no single method for determining the size or richness of an archaeobotanical sample. In this study I have opted for two measures, one being an extension of the other: abundance and average density. These values are given for each sample, where applicable, in Appendix 4.

'Abundance' here denotes the total quantity of whole, charred, standard plant parts in a given sample. So defined, abundance allows for the identification of numerically insufficient samples, i.e. those whose plant parts are too few to justify quantitative analysis, and whose composition may be unrepresentative and misleading (Jones 1991: 67). Abundance was used in tandem with soil volume (in litres) to calculate average density. 'Average density' is here defined as the quantity of whole, charred, standard plant parts per litre of sediment, for a given sample. It has been calculated only for those quorate samples with an abundance of at least 30 (Appendix 1: Parameter 7). In some excavation reports, no soil volumes were specified, or else a different unit of measure was employed (often weight in kilograms), such that average density could not be calculated for those samples.

Across the 736 samples in the project dataset, abundance ranges from one to 20,153 items. Of these samples, 239 are quorate (i.e. containing at least 30 items). Average density can be calculated for 199 of those 239 samples, and ranges from 0.1 to 1010 items per litre.

Average density can serve as a proxy for deposition rates, on the grounds that rapid deposition should result in a relatively dense concentration of plant remains (provided that there has been little later disturbance), whereas gradual, piecemeal deposition – or else later disturbance or redeposition – should result in a sparser distribution of plant remains in the sampled soil. Thus, average density can potentially shed light on a deposit's taphonomic history and so perhaps, by extension, elucidate its functional significance within its original plant economy (van der Veen 2007: 987, Table 6). For the purposes of this chapter, average density is also a better gauge than abundance of a sample's 'largeness' because it is directly comparable between samples which may differ significantly in soil volume.

Scales of production and processing

How do these data help us to address the question of increasing crop surpluses in Early and Mid Saxon farming? To reiterate the argument cited at the start of this chapter, van der Veen and Jones interpret 'large, accidentally charred grain-rich samples as representing large-scale production and/or consumption' (van der Veen and Jones 2006: 223). Those 44 free-threshing cereal samples which have been interpreted as representing fine-sieved products (FSP) or unsieved grain (USG), in the crop processing analyses above, may stand for the grain-rich samples in this model.

Two of these grain-rich samples are of Early Saxon date, six of Intermediate date, 31 of Mid Saxon date, and five of Generic date: already an indication that more charred grain-rich deposits were being produced in the 8th and 9th centuries than in the preceding three centuries. This distributional bias towards the Mid Saxon period is even more marked than that exhibited by the entire collection of samples in the dataset, which suggests that there is a genuine tendency for USG and FSP samples to be concentrated in the 8th and 9th centuries, over and above the tendency for more samples in general to be of Mid Saxon date (Table 10).

Table 10 – Chronological distribution of all samples compared with free-threshing grain-rich product (USG/FSP) samples.

phase	no. samples (all)	% samples (all)	no. USG/FSP samples	% USG/FSP samples
Early Saxon	207	28.1	2	4.5
Intermediate	151	20.5	6	13.6
Mid Saxon	321	43.6	31	70.5
Generic	57	7.7	5	11.4
Total	736	100.0	44	100.0

Are these grain-rich samples also large? Although there is no universal, objective yardstick for largeness, average density provides a useful comparative measure of the size of these samples, as discussed above. Average density is calculable for 36 of the 44 grain-rich samples, ranging from 0.8 to 154 items per litre. Of these 36 samples, two are of Early Saxon date, six of Intermediate date, 25 of Mid Saxon date, and three of Generic date. The 25 Mid Saxon samples not only outnumber the other samples; they also display markedly higher average densities (Figure 11). Of the 36 samples in question, the only ones with an average density higher than 20 items per litre are eight of Mid Saxon date, and two of Generic date (these two are from Harston Mill, a multi-period site, and could quite possibly be of Mid Saxon date).

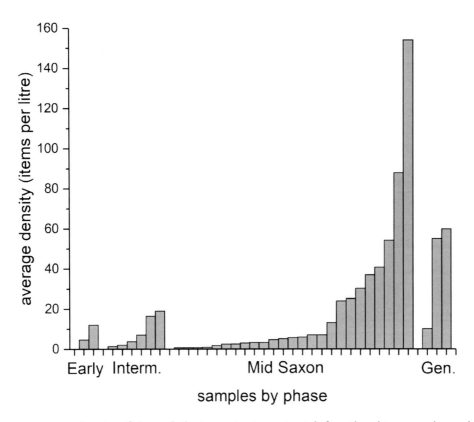

Figure 11 – Average density of charred plant remains in grain-rich free-threshing cereal samples, grouped chronologically.

The feature types from which these samples derive might also be exerting some influence on the average density of charred plant remains. There is a notable concentration of denser samples among pits (or wells) than among ditches, SFBs and other feature types. To some extent this goes hand-in-hand with the chronological trend, since pits in general are proportionally better represented among the Intermediate and Mid Saxon samples than among the Early Saxon samples, whereas the reverse is true of SFBs (Table 11; Figure 12).

Table 11 – Phased distribution of samples by feature type.

phase (# samples)	% ditch/gully	% hearth/oven	% other	% pit/well	% posthole	% SFB
Early Saxon (207)	5.3	1.4	7.7	17.4	14.0	54.1
Intermediate (151)	17.9	6.6	16.6	22.5	8.6	27.8
Mid Saxon (321)	24.9	1.6	33.0	26.2	3.7	10.6
Generic (57)	8.8	1.8	0.0	56.1	17.5	15.8
All samples (736)	**16.7**	**2.6**	**20.0**	**25.3**	**8.7**	**26.8**

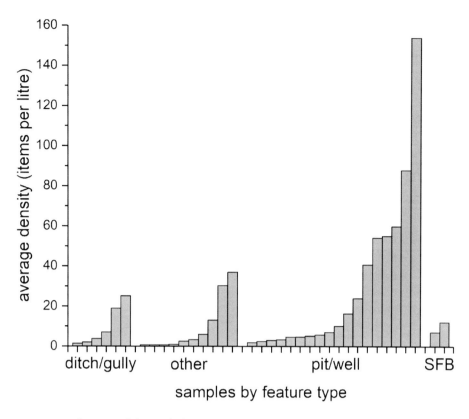

Figure 12 – Average density of charred plant remains in grain-rich free-threshing cereal samples, grouped according to parent feature type.

According to Tipper, most refuse in Early Saxon rural settlements was ultimately redeposited in the backfills of SFBs, having first accumulated in middens. From the Mid Saxon period onwards, pits are increasingly in evidence, although rubbish pits *sensu stricto* remain rare (Hamerow 2012: 94; Tipper 2004: 157–159). This general narrative sequence of refuse deposition is exemplified by the charred samples. For the Early Saxon period, more than 50% of samples derive from SFBs. This proportion declines steeply, progressively, through the Intermediate and Mid Saxon samples, while growing proportions of samples derive from ditches, the miscellaneous 'other' features and, less steeply, pits (Table 11). The increase in ditch-derived samples clearly parallels the growing occurrence of ditched features which characterises Anglo-Saxon settlements from the 7th century onwards. The pits, meanwhile, are of varied form, but none is closely comparable to the form of grain storage pit familiar from the Iron Age (Cunliffe 2005: 411–413). The single best represented context type in the 'other' category is the anthropogenic palaeosol at Mid Saxon Brandon, from which 34 grid square samples derive. 24 Mid Saxon samples from Ipswich, which also belong to the 'other' category, are likely to derive from either pits or ditches, but detailed information was not available at the time of writing.

Broadly, then, among Early Saxon samples there is a dual tendency towards (i) frequent deposition in SFB backfills and (ii) a lower average density of plant remains. Among the Mid Saxon samples, there is an opposite trend towards (i) higher average density of plant remains and (ii) more frequent deposition in ditches and pits. These two simultaneous developments – an increase in average density and a shift in dominant feature type – may be causally linked. If SFB backfills usually represent secondary or even tertiary deposition, as Tipper suggests, then one might well expect the charred plant remains therein to be of low average density, since any dense concentrations originally present in the primary deposits are likely to have been disturbed and diluted during redeposition (Tipper 2004: 157–159). Conversely, the high-density samples from Mid Saxon pits and ditches are arguably more likely to represent single, relatively rapid depositional events, such as batches of accidentally or deliberately burnt grain being promptly discarded, thus producing dense concentrations of charred plant remains (van der Veen 2007: 987, Table 6).

It would seem that such dense deposits are far more prevalent from the 8th century onwards, both in absolute terms and as a proportion of the entire dataset for the period. Arguably, the most straightforward interpretation of this development is that more material was being produced, processed and therefore burned in the Mid Saxon period, or at least that more material was being processed and stored in a centralised fashion.

However, the evidence cited above does not necessarily preclude the possibility that dense concentrations of charred grain were created prior to the 8th century, but were ultimately disposed of in a way that defies archaeological recognition, e.g. in middens that were subsequently reworked into cultivation plots, or as tertiary deposits in SFB backfills. However, it is striking that there are only three Early Saxon samples in the entire dataset with an average density exceeding 30 items per litre. Two of these, <AL1> and <SMB1> are anomalous because of their abundance of spelt glume bases. They are highly atypical and may better be considered as traces of a 'Very Late Roman' (post-410) phase of southern British agriculture, distinct from 'Early Saxon' farming. The third sample, <CRM3> from the Criminology Site in Cambridge, is more typical in that it is dominated by free-threshing cereal grains; it derives from the fill of an ovoid pit which displayed traces of burning, including burnt stones. It may conceivably represent a batch of cereals accidentally charred whilst being dried over a hearth, represented by the ovoid pit. If so, it would appear to be a unique example for this early period (Dodwell *et al.* 2004: 115, 119–120). It is also worth noting that a unique Early Saxon midden excavated at the royal centre at Lyminge (Kent) – one deposit of the period that we might expect to yield relatively high densities of charred plant remains – produced only sparse charred plant macrofossils even though charcoal was abundant (McKerracher 2015). On current evidence, then, dense deposits of charred plant remains in Early Saxon contexts do appear genuinely to be exceptionally rare.

It might be contended, however, that the observed chronological trends are an artefact of sampling strategies: that the study of certain Mid Saxon sites has benefited from the retrieval of more, and larger, samples. Against this argument it may be said that some Early Saxon sites, such as Eye, have in fact been subject to extensive sampling strategies, but few of the resultant samples have been deemed rich enough to warrant further analysis (Fryer 2008). In addition, since the measure of average density is calibrated by soil volume, the observed trends cannot be due simply to greater volumes of soil being taken from Mid Saxon contexts.

It might alternatively be contended, with particular regard to East Anglia, that the existence of Ipswich Ware as a diagnostic Mid Saxon pottery type renders deposits of that period more easily identifiable and datable, hence the apparently greater corpus of archaeobotanical material for the 8th and 9th centuries. However, even if Mid Saxon activity per se is easier to identify, this would not necessarily account for the greater average density of plant macrofossils within the samples of this period: a greater number of deposits need not mean a greater density of remains within those deposits. Rather, it would seem that Early Saxon charred plant remains are genuinely scarcer than their Mid Saxon counterparts, just as Early Saxon settlements often appear more ephemeral and transient than those of the Mid Saxon period. Intermediate material is typically more closely comparable to that of the Early Saxon period in terms of abundance and density, which implies that the major change in activity did not happen before the 8th century.

To summarise, there are exceptionally few samples of high density which can be dated to before the 8th century, to the extent that such samples may provisionally be considered to be, largely, a new phenomenon of the 8th to 9th centuries, potentially indicative of large-scale or centralised production and processing. It is instructive here to note that, among the 28 Mid Saxon samples with an average density greater than 30, four sites are particularly well represented: the Hutchison Site (five samples), Lake End Road (four), the Ashwell Site at West Fen Road, Ely (three), and Yarnton (five). The other sites, with one or two samples each, are Forbury House, Harston Mill, Lackford Bridge, Lot's Hole, Pennyland, Rosemary Lane, Brandon, Walpole St Andrew, and Ingleborough.

These 'high density' Mid Saxon sites occupy diverse different environments, but share other notable characteristics, such as high status and/or ecclesiastical associations (reflected by e.g. imported goods), and the occurrence of relatively rare crop taxa among the charred plant remains (e.g. opium poppy, glume wheats). It may therefore be conjectured that the proposed increase in the scale or centralisation of arable production and processing in the Mid Saxon period was particularly associated with high status and ecclesiastical establishments (McKerracher 2017).

Conversely, Ipswich, despite being represented by 35 Mid Saxon charred samples obtained via a wide-ranging sampling strategy (including the targeted sampling of very dense charred deposits discovered from other phases at the site), failed to produce any with average densities exceeding six items per litre for the Mid Saxon period. While this could to some extent be due to the extremely high soil volumes extracted (often more than 100 litres), which could have incorporated relatively small but still dense concentrations, it may be relevant that a similar tendency towards low densities of plant remains has also been observed among the Mid Saxon samples from Lundenwic, the emporium of Mid Saxon London (Davis in Malcolm *et al.* 2003: 290; Davis in Cowie *et al.* 2012: 300). One could conjecture that, as non-agricultural consumer populations, the inhabitants of the two emporia were not processing crops in bulk, and hence not giving rise to the theoretical scenario in which dense concentrations of charred grain would be produced and deposited. Rather, agricultural produce could have been acquired, processed and consumed on a domestic scale within individual households.

Glume wheat samples

The method described above – of calculating the relative proportions of grain, chaff and weed seed to identify crop processing stages – is not directly applicable to samples dominated by glume wheats, or those with a substantial proportion of glume wheats alongside free-threshing cereal remains, since glume wheats are processed in a different way: an additional stage, dehusking, is required after initial winnowing and sieving in order to extricate the grains from their encasing glumes, which must then be filtered out by further winnowing, sieving, and/or hand-sorting. Glume bases and rachis segments are also subject to different preservation biases. Hence, the ethnographically determined ratios of grain, rachis and weed seeds which are found by Jones (1990) to characterise different products and by-products may be compared only with those archaeological samples which are similarly dominated by free-threshing cereals. Nonetheless, the general botanical composition of the glume wheat-dominated samples (ratios of grain to glume bases to weed seeds) can be considered on a less formal basis. It is also possible to apply Jones' (1987) other method – based on the discriminant analysis of a sample's constituent weed seed types – to samples dominated by glume wheats, or those jointly dominated by glume wheats and free-threshing cereals.

The crop remains in <AL1>, <SMB1> and <BRT4> are dominated by glume wheats (specifically spelt wheat, where identifiable). Given the degree of dominance of glume wheats in these samples, the grains and glume bases of indeterminate cereals and indeterminate wheats have been counted as glume wheats for the purposes of the following calculations.

The crop content of these samples suggests that they do not represent cleaned grain products, since all three have relatively high proportions of glume bases and weed seeds (Table 12). Meanwhile, according to the discriminant analysis, <SMB1> and <BRT4> are likely to represent the by-products of fine sieving, while <AL1> is classified as the by-product of winnowing (Table 13; Figure 13).

Table 12 - Relative proportions of grain, glume base and weed seed in samples dominated by glume wheats.

sample	% grain	% glume base	% weed seed	total
AL1	21.0	42.3	36.7	1048
BRT4	34.3	48.3	17.4	385
SMB1	7.1	48.7	44.2	1179

Table 13 - Crop processing analysis of glume wheat samples by discriminant analysis of weed seed types (discriminant functions rounded to three decimal places).

sample	interpretation	Function 1	Function 2
AL1	WBP	0.708	1.878
BRT4	FSBP	-1.994	-0.943
SMB1	FSBP	-2.937	-3.946

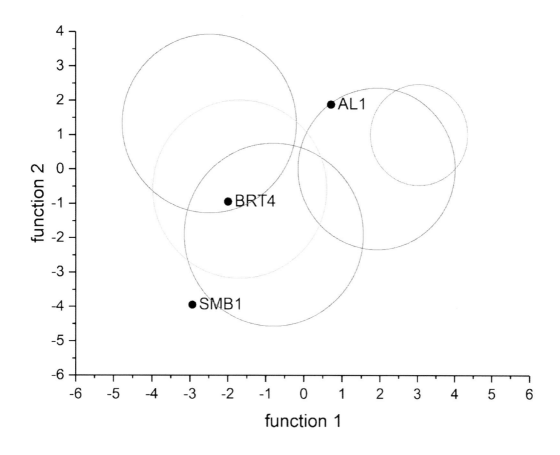

Figure 13 – Discriminant analysis scattergraph of glume wheat samples.

The results of these two approaches (by basic composition and discriminant analysis) can be reconciled as follows. The 'small, free, heavy' seeds and high ratio of glume bases to grain in <SMB1> are both consistent with this sample representing a fine-sieving by-product. Sample <BRT4>, by contrast, could represent the product of coarse-sieving, i.e. spikelets charred prior to fine-sieving: hence the discriminant analysis classification roughly in between the FSP and FSBP groups, and the similar proportions of grains and glume bases (some of the latter preserved as whole spikelet forks). Such spikelets could conceivably have been burned whilst being parched for dehusking – although, in this instance, whole spikelets do not seem to have been preserved intact (van der Veen 1989: 304). Finally, Pelling suggests that <AL1> represents a mixture of different processing by-products, including 'cumins', the by-product of rubbing parched and malted grains, as represented chiefly by glume bases and germinated grains (Pelling in Booth *et al.* 2001: 422). More than half of the spelt grains in <AL1> do show signs of having germinated. The discriminant analysis classification (winnowing by-product) is not consistent with the proportions of grains, glume bases and weed seeds in this sample, since the winnowing by-product of a glume wheat ought not to contain so many grains and glume bases. The sample's contents thus appear to be consistent with the mixed origins posited by Pelling.

In terms of abundance, as defined earlier in this chapter, all three samples comfortably exceed the quorum for numerical sufficiency, i.e. 30 items (Appendix 1: Parameter 7): there are 1075 items in <AL1>, 440 in <BRT4>, and 1239 in <SMB1>. While density could not be calculated for <BRT4> because soil volume data were not available, the other two samples both have a high average density, much the highest of any samples dated to the 5th and 6th centuries: 215 items per litre in <AL1>, 112.6 items per litre in <SMB1>. So, if these two samples do indeed date from the 5th or 6th century, they appear to represent a phenomenon distinct from their contemporaries in the dataset: dominated by spelt wheat, and very dense. As suggested above, they might better be considered as the remains of a 'Very Late Roman' mode of agriculture persisting into the 5th century.

Mixed cereal samples

As established earlier in this chapter, the crop contents of 15 samples are jointly dominated by free-threshing cereals and glume wheats. These samples can be subjected to the discriminant analysis of weed seed types (Jones 1987), following the methods and criteria outlined above. Given the mixture of cereal crop types in these samples, however, the relative proportions of grain, chaff and weed seed are less straightforward to interpret for these samples than for those dominated by free-threshing cereals alone.

In nine of these 15 samples, cereal grains constitute more than 80% of the combined grain, glume base, rachis segment and weed seed total (Table 14): these nine include all seven of the samples from Harston Mill, one from Ely (Chiefs St), and one from Yarnton. Only four of these samples were eligible for the discriminant analysis of weed seed types (Table 15; Figure 14), and all of them were thus classified as fine-sieving by-products. This classification is not compatible with their grain-rich composition and therefore suggests that these samples may not each represent a singular, coherent deposit. Two samples, <ENB2> and <Y4>, were jointly dominated by cereal grains and weed seeds (Table 14). The former was classified by the discriminant analysis as a winnowing by-product, which is not compatible with its relatively grain-rich contents, while the latter was classified as a fine-sieving by-product, which could be compatible with this sample representing unsieved grain (Table 15; Figure 14). In the remaining six samples, weed seeds constitute more than 50% of the combined total number of items (Table 14), and the discriminant analysis classifies three of them as fine-sieving by-products and one as a fine-sieved product (Table 15; Figure 14). The by-product classifications might seem compatible with the high proportions of weed seeds in these samples, but they are not so compatible with the relatively high (>6%) proportions of rachis. Strictly speaking, therefore, only <WST1> from West Stow can be seen as a plausibly discrete fine-sieving by-product sample.

Table 14 - Relative proportions of grain, chaff and weed seed in mixed cereal samples.

sample	phase	% grain	% rachis	% glume base	% weed seed	total items
CSE2	Mid Saxon	92.0	0.0	0.0	8.0	138
ENB2	Early Saxon	56.0	2.2	1.5	40.3	273
HAM2	Mid Saxon	99.6	0.0	0.0	0.4	562
HAM26	Mid Saxon	96.4	0.0	0.0	3.6	829
HAM3	Generic	96.1	0.0	0.0	3.9	77
HAM32	Mid Saxon	85.6	0.0	0.0	14.4	382
HAM36	Generic	97.4	0.0	0.0	2.6	230
HAM4	Generic	94.2	0.0	0.0	5.8	295
HAM6-7	Generic	95.2	0.0	0.0	4.8	331
LW2-3	Mid Saxon	25.8	8.5	13.0	52.6	1533
LW4	Mid Saxon	21.7	17.1	5.1	56.0	2825
LW7	Mid Saxon	19.4	6.8	3.8	70.0	1458
WST1	Early Saxon	10.5	0.2	2.9	86.3	2085
Y25	Mid Saxon	86.1	0.0	6.6	7.3	287
Y4	Early Saxon	43.2	0.0	16.2	40.5	74

Table 15 - Crop processing analysis of mixed cereal samples by discriminant analysis of weed seed types (discriminant functions rounded to three decimal places).

sample	interpretation	Function 1	Function 2
ENB2	WBP	0.084	0.166
HAM26	FSBP	-2.899	-2.236
HAM32	FSBP	-2.430	-1.193
HAM6-7	FSBP	-3.165	-1.212
LW2-3	FSP	-1.979	0.467
LW4	FSBP	-1.579	-2.897
LW7	FSBP	-1.434	-3.093
WST1	FSBP	-1.469	-2.762
Y25	FSBP	-1.059	-4.943
Y4	FSBP	-2.386	-2.321

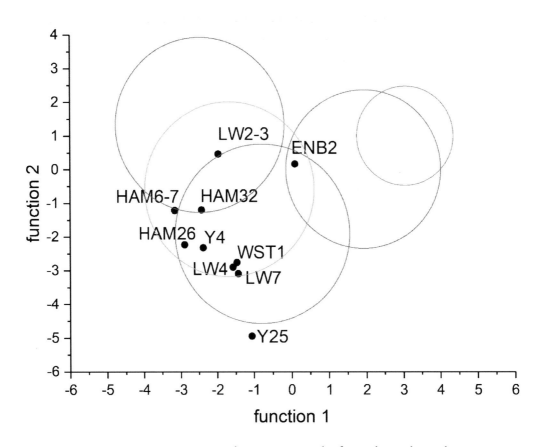

Figure 14 – Discriminant analysis scattergraph of mixed cereal samples.

In total, therefore, these comparative analyses result in no more than two of the 15 samples (<Y4> and <WST1>, both Early Saxon) being assigned to a particular crop processing stage. It is, of course, unsurprising that samples with a mixture of cereal types should also represent a mixture of crop processing products and by-products, since the different cereal types have different processing requirements. On the other hand, it is fair to note that in five of the remaining samples, cereal grain constitutes more than 90% of the total number of items and the corresponding lack of weed seeds prohibits discriminant analysis (Table 14). These five might therefore provisionally be considered potentially to represent fine-sieved products.

Average density could be calculated for 14 of the 15 mixed cereal samples (since no volume data were available for <CSE2>), and ranges from 0.5 to 78 items per litre (Table 16). It was observed earlier in this chapter that the only free-threshing grain-rich products samples with an average density higher than 20 items per litre were of Mid Saxon date, and that is also true of these 14 mixed cereal samples, six of which have a density greater than 20, all dating from the Mid Saxon phase.

Table 16 – Abundance and average density of charred plant remains in mixed cereal samples.

sample	phase	abundance	average density
CSE2	Mid Saxon	54	-
ENB2	Early Saxon	122	12.2
HAM2	Mid Saxon	569	28.5
HAM26	Mid Saxon	841	42.1
HAM3	Generic	77	3.9
HAM32	Mid Saxon	383	19.2
HAM36	Generic	230	11.5
HAM4	Generic	118	5.9
HAM6-7	Generic	334	8.4
LW2-3	Mid Saxon	1554	51.8
LW4	Mid Saxon	2825	70.6
LW7	Mid Saxon	1471	36.8
WST1	Early Saxon	2085	0.5
Y25	Mid Saxon	390	78.0
Y4	Early Saxon	75	3.8

Summary

The aim of this chapter has been to interpret the samples in the project dataset as artefacts of early medieval crop production and processing activities. Simple statistical assessments have established that, where calculable, most samples are dominated by the remains of free-threshing cereals. Those three which are unusually dominated by spelt wheat could perhaps be interpreted in the context of a long twilight of Romano-British crop husbandry. Through the combined application of a ratio-based analysis and the discriminant analysis of weed seed types, I have also attempted to classify samples in terms of which processing products and by-products they may represent, and so identified a core set of 44 grain-rich free-threshing product samples, representing fine-sieved products or unsieved grain (following Jones 1987; 1990).

The purpose of these exercises has been twofold. First, it defines those 44 samples as a group of comparable artefacts with coherent botanical contents, whose constituent crop and weed species will be analysed in the following two chapters. Second, in tandem with the calculation of average densities of charred plant remains, it offers a window on the growth of surplus crop handling in the Mid Saxon period. While the emergent pattern might be due in part to a shift in depositional practices around the 7th and 8th centuries – away from deposition or redeposition in SFB backfills and towards deposition in pits and ditches – it is possible that it nonetheless represents a real growth in crop surpluses in the Mid Saxon period, especially perhaps at sites with high-status associations. The question remains, however, as to what specific crops were contributing to this supposed growth in surplus production and processing?

Chapter 5: Counting the Crops

This chapter investigates patterns in the range and relative importance of the cereal crops in Early and Mid Saxon agriculture, through a combination of semi-quantitative presence analyses and fully quantitative analyses of relative abundance. There are some current hypotheses surrounding crop-biological innovations in this period that can be tested against the results obtained in this chapter. It has variously been argued that bread wheat supplanted spelt at the start of the Anglo-Saxon period; that bread wheat overtook barley as the most important cereal crop in the Mid Saxon period; that rye and oat became more important crops in the Mid Saxon period; and that emmer enjoyed a localised resurgence at around the same time. How well do these models fit with the project dataset?

A large proportion of Chapter 4 was devoted to the classification of samples in terms of dominant crop types and crop processing stages, partly with the aim of demarcating sets of comparable 'artefacts' so that in this chapter we can compare like with like when analysing the proportions of different cereal grains represented in these samples. The presence analyses in this chapter, however, are conducted on a much more inclusive basis, embracing all samples regardless of crop type or processing stage. This inclusivity is justified, I would argue, by the complementary objectives of the semi-quantitative and fully quantitative analyses. The semi-quantitative work aims to gauge the prevalence of species in the archaeobotanical record, i.e. whether they are ubiquitous, common or rare. The fully quantitative work aims to assess the relative abundance of those species in crop-processing products, as a proxy for their abundance or scarcity in Anglo-Saxon harvests. In these terms, it would be possible for a species simultaneously to be both rare and abundant.

The importance of importance

It is all too easy to debate the importance of different cereal crops in past agricultural regimes without defining what is meant by 'importance', and how it can be measured archaeobotanically. There is no universal answer to that problem. Possible definitions of importance – by no means comprehensive or mutually exclusive – include the following concepts.

- **Dietary importance:** how much of a community's calorific intake is provided by the crop?
- **Social importance:** how high is the crop's cultural status?
- **Economic importance:** how central is a crop to a community's production and exchange activities?
- **Agronomic importance:** how central is a crop to the productivity of arable farming regimes?

Since charred plant remains are the (by-)products of crop processing, they are not apt to shine any direct light on dietary or social importance, beyond illustrating what crops were available at which sites. Economic – and, especially, agronomic – importance is more likely to be elucidated directly by the project dataset, but it is still necessary to parse out the different aspects of economic/agronomic importance that can feasibly be investigated by quantitative and semi-quantitative archaeobotanical methods. In this context, I propose to assess not the economic importance of crops *per se*, but their prevalence, frequency of use, and relative productivity.

By prevalence, I mean simply the breadth of distribution for a given region or period: how widespread each crop is in the archaeobotanical record, as a proxy for how widespread it was in Early and Mid Saxon England. This can be investigated straightforwardly by presence analysis, i.e. the calculation of the proportion of assemblages in a given period or region, in which a particular crop occurs.

By frequency of use, I mean how regularly a crop was handled – grown, harvested, threshed, stored – in a particular region, or during a particular period. A proxy for this characteristic can again be found in presence analysis, this time based on individual, independent samples: the calculation of the proportion of samples in a given region or period, in which a particular crop taxon occurs. The justification for using this measure as a proxy for handling-frequency is the premise that charred crop deposits represent the visible tip of a sunken crop processing iceberg: the more often a crop was harvested and processed, the more likely it is to have been preserved and recovered in charred deposits.

Finally, by relative productivity, I mean how much a crop contributed to the total harvested cereal goods in a given region or period, relative to the other cereals grown in that region or period. This must be a relative measure, since there is no obvious way of inferring absolute yields from archaeobotanical remains which are unlikely, in most cases, to derive from primary storage contexts (see Chapter 4). The relative abundance of the grains of different cereals in individual samples will here be used as a proxy for relative productivity. This strategy needs some explanation and clarification.

It has been established in Chapter 4 that free-threshing cereals constitute the dominant crop type in the project dataset, and that a significant subset of the samples thus dominated can be defined as grain-rich products (USG and FSP classifications in Chapter 4). The functional integrity of these 44 samples as discrete products of cereal processing – such that each may plausibly represent a cache of cereals processed and perhaps also grown together – means that the relative abundance of cereal grains in each sample offers a snapshot of the relative productivity of those crops in a past processing event. Since we cannot determine how representative an individual sample might be of its parent economy, it is best for this kind of analysis to cast its net widely over several settlements, so that the broader picture of a region or period may include as many samples as possible. Put simply, since individual samples might not be representative, they are most usefully studied in aggregate. Of course, the necessary restriction of this analysis to the 44 USG and FSP samples, as a set of comparable artefacts, means that it can only address questions pertaining to free-threshing cereals. Questions relating to the relative productivity of glume wheats will therefore be addressed in a separate section of this chapter.

It should be stressed once again that the concepts employed here of prevalence, usage-frequency and relative productivity are not meant to be synonyms for importance. Rather, they are meant to describe different facets of the more complex properties of economic or agronomic importance, so that we can describe more precisely the patterns that emerge from the archaeobotanical data. Prevalence, frequency and relative productivity are theoretically independent properties and could therefore follow different trajectories at the same time. For instance, in principle a crop could become less prevalent over time but simultaneously increase in relative productivity at those sites which continued to cultivate it.

Prevalence

A simple presence analysis was used to calculate prevalence. For a defined subset of site assemblages in the project dataset – such as those assemblages from a given region or period – the number of assemblages in which a taxon appears was calculated as a percentage of the total number of assemblages in that subset.

I have conducted this kind of presence analysis on a chronological basis by grouping assemblages by phase, and calculating the prevalence of each crop taxon for each phase: Early Saxon (5th to 6th centuries), Intermediate (7th to early 9th centuries), Mid Saxon (8th to later 9th centuries), and Generic (5th to 9th centuries). The minimum number of units required to perform presence analysis is another key parameter that must be defined. Here I have opted for a minimum of ten units (assemblages or samples) in a region or period, before presence analysis may be carried out (Appendix 1: Parameter 8).

In theory, a geographical dimension could be explored by grouping assemblages by National Character Area (NCA), and calculating presence for each NCA (see Chapter 2). In practice, however, most NCAs contain too few assemblages to allow meaningful percentages to be calculated. Out of a total of 24 NCAs, only three contain more than ten assemblages apiece; 11 of them contain only one assemblage each (Table 3). Analysing prevalence across only three out of 24 NCAs would be a poor basis for inferring geographical patterns, so in this instance presence analysis has not been performed on a geographical basis.

Presence analyses can apply at different taxonomic levels. For instance, it can be calculated individually for free-threshing wheat grains, indeterminate 'Triticum sp.' grains, spelt grains, emmer grains, and ambiguous 'emmer/spelt' grains. But it is also useful to calculate presence for composite taxon groups, such as all glume wheats at once (encompassing emmer, spelt, and ambiguous emmer/spelt), or indeed for all wheats at once. Such figures must be calculated independently because presence values are not additive. Emmer and spelt grains could, for instance, both occur in the same sample, so simply adding together their presence values could result in redundant double-counting. The results of the chronology-based presence analysis are displayed in Table 17, including composite taxon groups preceded by an asterisk (*).

Table 17 – Phased presence analysis of cereal taxa by assemblages.

taxa	% assemblages where present			
	Early Saxon	Intermediate	Mid Saxon	Generic
* Cereals (all)	100.0	100.0	97.4	100.0
Cereal indet.	86.5	91.3	92.1	92.3
Hordeum L.	94.6	95.7	92.1	84.6
* Wheats (all)	94.6	95.7	94.7	84.6
Triticum L. indet.	78.4	69.6	78.9	61.5
Triticum L. (free-threshing)	51.4	78.3	71.1	53.8
* Glume wheats (all)	54.1	39.1	44.7	30.8
Triticum dicoccum Schübl./*spelta* L.	29.7	21.7	28.9	23.1
Triticum spelta L.	40.5	17.4	28.9	23.1
Triticum dicoccum Schübl.	8.1	13.0	15.8	7.7
* Oats (*sativa* + indet.)	54.1	73.9	73.7	69.2
Avena L. indet.	51.4	69.6	71.1	69.2
Avena sativa L.	2.7	4.3	7.9	0.0
Secale cereale L.	24.3	56.5	65.8	53.8
Total number of assemblages (n = 111)	37	23	38	13

What chronological patterns emerge from this analysis? Cereals in general are almost entirely ubiquitous, and indeterminate cereals nearly so. Barley (*Hordeum* L.) and wheat (*Triticum* L.) are by a wide margin the most prevalent cereal genera throughout the whole Early to Mid Saxon period, appearing in over 90% of assemblages of Early, Intermediate, Mid Saxon and Generic date. Differences between phases are negligible, amounting to a rise or fall of two or three percentage points at the very most, so it would be fair to say that both barley and wheat are prevalent to the point of ubiquity among the entirety of assemblages. There is no evidence that either of these crops became more or less widespread between the 5th and 9th centuries (Figure 15).

Within the *Triticum* genus, however, there are some distinct patterns among the different kinds of wheat in the dataset. Perhaps surprisingly, free-threshing and glume wheats are practically as widespread as each other in the Early Saxon phase, the prevalence of glume wheats being accounted for in the main by *Triticum spelta*. The prevalence of free-threshing wheat in the Intermediate and Mid Saxon phases, i.e. between the 7th and 9th centuries, is 20 to 30 percentage points higher than in the Early Saxon phase, with a modest contraction in the prevalence of free-threshing wheat between the Intermediate and Mid Saxon phases (Figure 16).

While free-threshing wheat thus becomes more widespread – or at least more identifiable – from the 7th century onwards, the prevalence of the glume wheats is ten to 15 percentages points lower in the Intermediate and Mid Saxon phases than in the preceding 5th and 6th centuries. Emmer wheat remains a restricted presence among the glume wheats, never becoming more widespread than spelt, but its prevalence does rise marginally over time, while that of spelt contracts by more than 20 percentage points in the Intermediate phase before expanding again a little in the 8th and 9th centuries (Figure 17).

Oat (*Avena* L., including *A. sativa* L.) and rye (*Secale cereale* L.) are much less widespread than wheat and barley in every phase, but both nonetheless become markedly more prevalent over time. Interestingly, in the Early Saxon phase, oat is practically as prevalent as free-threshing and glume wheats. Presence values for both oat and rye show the biggest rise in the Intermediate phase, i.e. from the 7th century onwards, and this leap is particularly pronounced for rye, whose prevalence more than doubles between the Early and Intermediate phases. Both oat and rye are much more prevalent than emmer wheat, and rise to become more widespread than glume wheats in general during the 7th to 9th centuries (Figure 18).

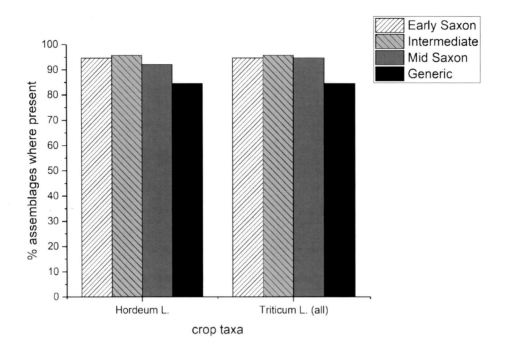

Figure 15 – Phased presence analysis of barley and wheat remains, by assemblages.

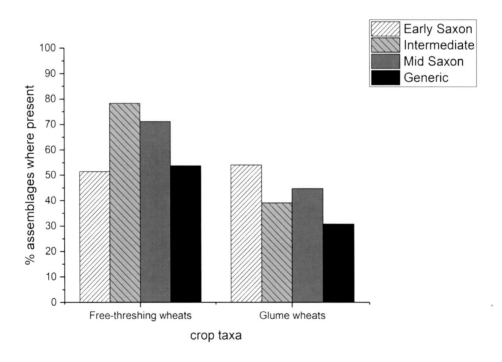

Figure 16 – Phased presence analysis of free-threshing and glume wheat remains, by assemblages.

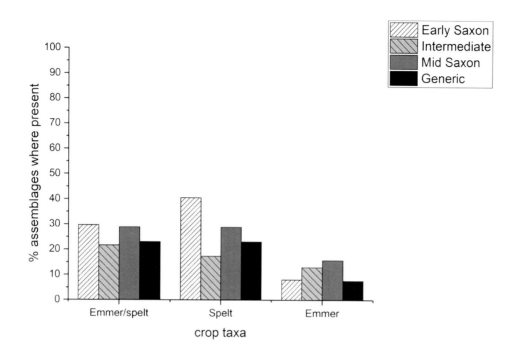

Figure 17 – Phased presence analysis of different glume wheat remains, by assemblages.

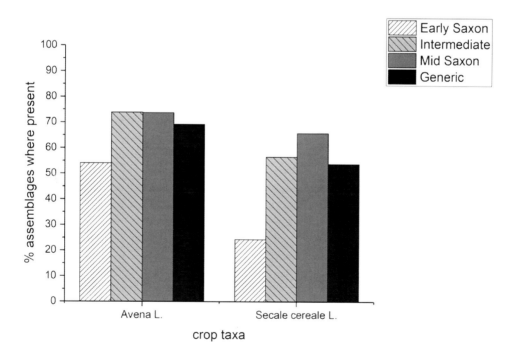

Figure 18 – Phased presence analysis of oat and rye remains, by assemblages.

Frequency of use

The method here is exactly the same as that used for prevalence, except the unit for enumeration is the individual independent sample, rather than the assemblage. Thus for a defined subset of samples – representing a given region or period – the number of samples in which a taxon appears is calculated as a percentage of the total number of samples in that subset.

As described above for the 'prevalence' methodology, presence values for composite taxonomic groups (e.g. glume wheats in aggregate) must be calculated independently so as to avoid misleading double-counting. The composite taxonomic groups are preceded by an asterisk (*) in Table 18.

Just as cereals in general are practically ubiquitous among assemblages, so are they also among samples. The usage-frequency of barley increases by around ten percentage points from around the 7th century, then changes little between the Intermediate and Mid Saxon phases. The usage-frequency of the wheat genus, as a whole, follows a similar trajectory, but with a spike in the Intermediate phase that could indicate more frequent usage of wheat than barley around the 7th century (Figure 19). Within the *Triticum* genus, free-threshing wheat appears to be used much more frequently in the Intermediate and Mid Saxon phases than earlier. The values for glume wheats, individually and collectively, change little over time, declining very slightly and always remaining lower than those for free-threshing wheat (Figure 20). The modest overall reduction in the use of spelt and the even subtler rise in the use of emmer are arguably too small to be very informative. The relatively high presence of spelt among samples of Generic date is accounted for largely by Harston Mill (Figure 21). More marked are the patterns for oat and especially rye, both of which were used progressively more frequently through the Intermediate and Mid Saxon phases, surpassing the usage-frequency of the glume wheats from the 7th century onwards (Figure 22).

Table 18 – Phased presence analysis of cereal taxa by samples.

taxa	% samples where present			
	Early Saxon	Intermediate	Mid Saxon	Generic
* Cereals (all)	91.8	99.3	96.3	98.2
Cereal indet.	74.9	83.4	89.1	86.0
Hordeum L.	57.5	68.9	67.6	80.7
* Wheats (all)	59.4	77.5	64.8	77.2
Triticum L. indet.	45.9	47.0	37.7	56.1
Triticum L. (free-threshing)	20.8	53.0	47.7	38.6
* Glume wheats (all)	19.3	11.9	16.2	33.3
Triticum dicoccum Schübl./*spelta* L.	9.7	6.6	10.3	10.5
Triticum spelta L.	11.1	3.3	6.9	26.3
Triticum dicoccum Schübl.	1.4	2.6	4.4	3.5
* Oats (*sativa* + indet.)	22.2	36.4	41.1	45.6
Avena L. indet.	21.3	35.8	40.8	45.6
Avena sativa L.	1.0	0.7	1.2	0.0
Secale cereale L.	7.7	25.2	40.2	24.6
Total number of samples (n = 736)	**207**	**151**	**321**	**57**

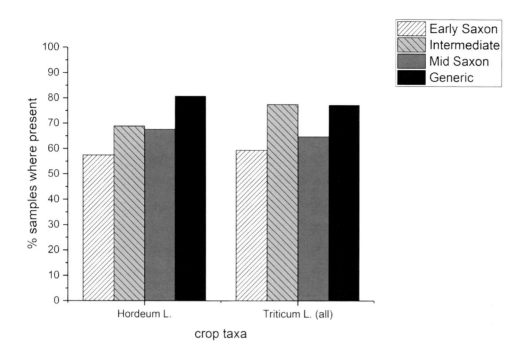

Figure 19 – Phased presence analysis of barley and wheat remains, by samples.

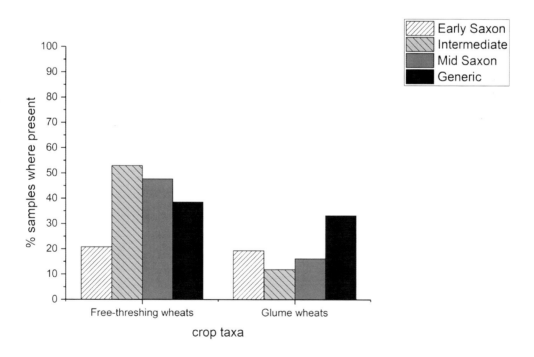

Figure 20 – Phased presence analysis of free-threshing and glume wheat remains, by samples.

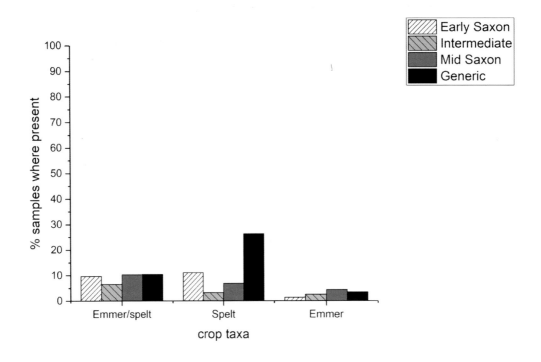

Figure 21 – Phased presence analysis of different glume wheat remains, by samples.

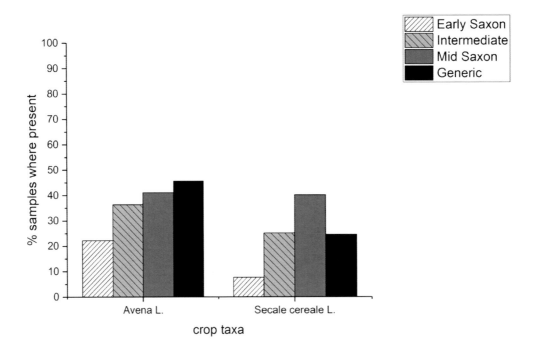

Figure 22 – Phased presence analysis of oat and rye remains, by samples.

I have also calculated usage-frequency for the different cereal taxa in terms of National Character Areas (NCAs), omitting those 11 NCAs which contain ten or fewer samples, in accordance with the parameter specified above (Appendix 1: Parameter 8). This leaves 13 NCAs for which usage-frequency values can be calculated. The results are shown in Table 19. These data shed light on geographical patterns in the cultivation and handling of the different crop taxa, and may usefully be illustrated cartographically.

In brief, barley is used noticeably less frequently in the Bedfordshire and Cambridgeshire Claylands, the Breckland, Salisbury Plain and the West Wiltshire Downs, and the South Norfolk and High Suffolk Claylands – together forming a diverse and heterogeneous group of clayey, sandy and calcareous upland terrains (Figure 23). Wheats (including indeterminate, free-threshing and glume wheats) register lower usage-frequency around the South Norfolk and High Suffolk Claylands, the Suffolk Coast and Heaths, the Fens, and especially the Breckland (Figure 24). A similar pattern – but including also the Cotswolds and North West Norfolk as areas of lower frequency – is shown specifically by free-threshing wheat (Figure 25).

Spelt is used relatively frequently around the Cotswolds, North West Norfolk and especially the East Anglian Chalk (a pattern influenced largely by Harston Mill), while emmer registers its highest values again around the Cotswolds, the Suffolk Coast and Heaths, but above all in the Thames Valley, the latter being influenced primarily by Lake End Road (Figures 26 and 27).

The usage-frequency of oat is very variable between the different NCAs, but is noticeably higher around North West Norfolk and the Thames Valley, and particularly low around the Avon Vale and especially Salisbury Plain and the West Wiltshire Downs (Figure 28). Values for rye are similarly varied, but highest around the Breckland, North West Norfolk, the South Suffolk and North Essex Claylands, and Thames Valley; and lowest around the Avon vale, Cotswolds, Salisbury Plain and West Wiltshire Downs, and the South Norfolk and High Suffolk Claylands (Figure 29). The concentration around the South Suffolk and North Essex Claylands is particularly influenced by rye's frequent presence among the samples at Ipswich.

Table 19 – Presence analysis of cereal taxa by samples, in terms of National Character Areas (including only NCAs with more than ten samples).

taxa	Avon Vale	Bedfordshire and Cambridgeshire Claylands	Breckland	Cotswolds	East Anglian Chalk	North West Norfolk	Salisbury Plain and West Wiltshire Downs	South Norfolk and High Suffolk Claylands	South Suffolk and North Essex Clayland	Suffolk Coast and Heaths	Thames Valley	The Fens	Upper Thames Clay Vales
*Cereals (all)	100.0	96.5	94.5	100.0	100.0	100.0	100.0	85.6	96.6	93.3	100.0	98.2	97.6
Cereal indet.	75.0	85.1	84.6	90.9	89.1	100.0	78.8	71.1	86.2	66.7	95.2	96.4	83.1
Hordeum L.	90.0	59.6	49.5	81.8	93.5	92.3	39.4	40.0	75.9	66.7	100.0	83.9	74.2
*Wheats (all)	85.0	72.8	37.4	100.0	91.3	76.9	60.6	47.8	89.7	46.7	100.0	51.8	79.8
Triticum L. indet.	65.0	57.0	26.4	100.0	60.9	76.9	6.1	41.1	29.3	20.0	85.7	16.1	55.6
Triticum L. (free-threshing)	55.0	43.9	9.9	27.3	52.2	23.1	57.6	6.7	63.8	20.0	100.0	46.4	57.3
*Glume wheats	0.0	14.0	8.8	36.4	65.2	23.1	0.0	7.8	6.9	20.0	71.4	16.1	19.4
Triticum dicoccum Schübl./spelta L.	0.0	11.4	4.4	9.1	8.7	0.0	0.0	0.0	0.0	20.0	71.4	14.3	16.1
Triticum spelta L.	0.0	5.3	4.4	18.2	58.7	23.1	0.0	7.8	5.2	6.7	4.8	0.0	5.6
Triticum dicoccum Schübl.	0.0	1.8	0.0	18.2	6.5	0.0	0.0	0.0	1.7	13.3	42.9	1.8	1.6
*Oats (sativa + indet.)	15.0	31.6	22.0	63.6	52.2	84.6	9.1	18.9	60.3	20.0	100.0	23.2	40.3
Avena L. indet.	15.0	31.6	22.0	63.6	52.2	84.6	9.1	18.9	58.6	20.0	95.2	23.2	40.3
Avena sativa L.	0.0	0.0	0.0	0.0	0.0	0.0	0.0	0.0	3.4	0.0	14.3	0.0	0.0
Secale cereale L.	5.0	21.9	48.4	9.1	30.4	53.8	6.1	6.7	63.8	33.3	61.9	17.9	19.4
Total number of samples (n = 692)	20	114	91	11	46	13	33	90	58	15	21	56	124

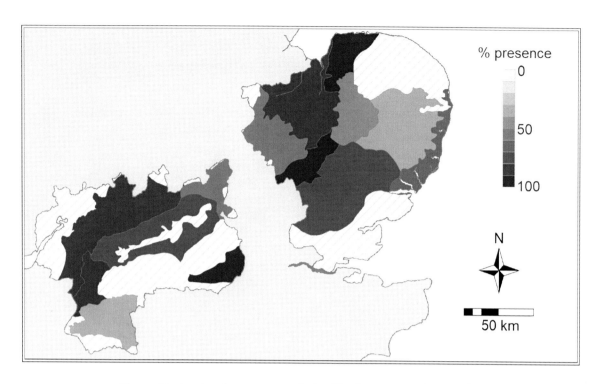

Figure 23 – Regional presence analysis of barley remains, by samples.

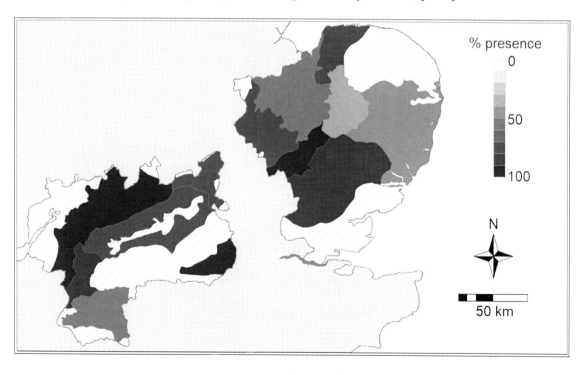

Figure 24 – Regional presence analysis of wheat remains, by samples.

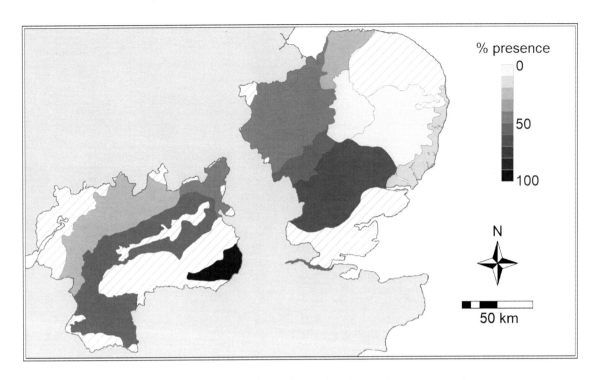

Figure 25 – Regional presence analysis of free-threshing wheat remains, by samples.

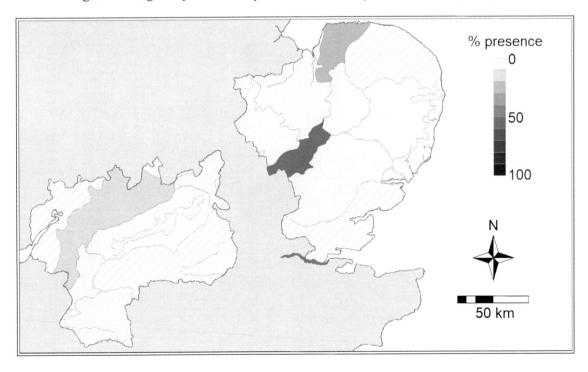

Figure 26 – Regional presence analysis of spelt remains, by samples.

Figure 27 – Regional presence analysis of emmer remains, by samples.

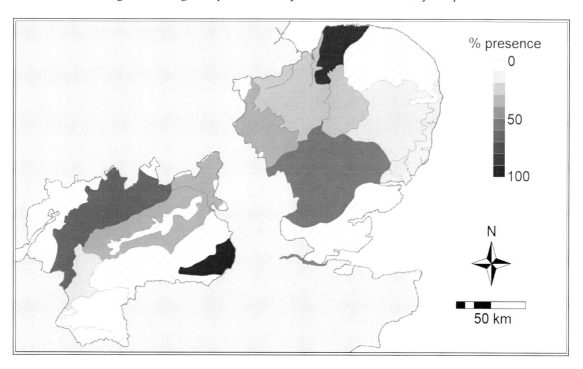

Figure 28 – Regional presence analysis of oat remains, by samples.

Figure 29 – Regional presence analysis of rye remains, by samples.

Relative productivity: free-threshing cereals

The calculations of prevalence and frequency of use, as executed above, are both forms of presence analysis, and thus take no account of the quantified cereal remains within individual samples. Hence they provide no indication of how much each crop contributed to the arable yields of the different sites, regions and periods, i.e. their relative productivity. The relative proportions of quantified cereal remains contained within the different samples may serve as a proxy for relative productivity, subject to certain constraints and assumptions. For instance, there must be a single unit of quantification, to ensure fair comparisons between taxa and between samples: grains, being much more common than chaff in the project dataset, are the best such unit for quantification.

This section focuses exclusively on the relative productivity of the free-threshing cereals within those samples dominated by free-threshing cereal remains (see Chapter 4), because these cereals share a common processing sequence and their grains may plausibly therefore have been grown, stored and/or processed together. As a further consequence of these samples being dominated by free-threshing cereals (≥80%), all indeterminate '*Triticum* sp.' grains have been amalgamated with the positively-identified free-threshing wheat grains. The few remaining glume wheat grains, if any, have been omitted as potential contaminants.

The method first requires that we add together the grain counts for each free-threshing cereal taxon – barley, wheat, oat and rye – in each sample. Only those samples which contain at least 30 grains of the four cereals in total are eligible for the calculation of relative productivity (Appendix 1: Parameter 9). The percentage of this total contributed by each of the four cereals is then calculated for each sample in turn. To maximise the number of samples available for this kind of analysis, one could include all samples dominated by free-threshing cereals, irrespective of the crop processing analyses conducted in Chapter 4. Such is the basis for the calculations used in Chapter 5 of the companion volume, *Farming Transformed* (McKerracher 2018: 103–111).

However, a more rigorous approach would include only those samples which represent comparable stages in the crop processing sequence: in this case, the 44 samples classed in Chapter 4 as grain-rich products, i.e. USG and FSP samples. A reason for restricting analysis to this subset of samples, at the cost of a much reduced dataset, is that each of these samples can more plausibly be thought to represent a single processing event. In other words, because these samples are less likely to represent mixed deposits, the cereals represented therein are more likely to have genuinely co-occurred in these proportions at their parent sites at the time of deposition, perhaps even within their parent harvests.

The overall results for this relative productivity analysis of the 44 free-threshing product samples are presented in Table 20. A significant disadvantage of this more rigorous sample selection policy is that it leads to the exclusion of most of the Early Saxon samples, and an overwhelming predominance of Mid Saxon samples (Table 21). While this imbalance is likely symptomatic of a general increase in cereal production and processing from 7th and 8th centuries onwards (as argued in Chapter 4), it nonetheless unfortunately restricts our ability to discern diachronic patterns in relative productivity. Meanwhile, eight National Character Areas are represented by these 44 samples, and among these there is a heavy bias towards the South Suffolk and North Essex Claylands (Table 22). This imbalance is due to the disproportionately strong representation of Ipswich, which contributes 16 of the 17 samples from this NCA.

Given the somewhat narrow scope of this more 'exclusive' version of the analysis, in comparison with the more 'inclusive' version presented in *Farming Transformed*, it may be most useful to consider the two sets of results in tandem, combining their respective strengths of analytical rigour and wide scope. In the following discussion, therefore, I have revisited the observations set out in *Farming Transformed* and considered how well they are supported by the 'exclusive' results derived in this chapter.

To aid direct comparison of the inclusive and exclusive results, I have visualised the data in this chapter in the same way as in *Farming Transformed*. First, the data are plotted as bar charts, with each bar representing a sample, and the y-axis measuring the percentage of grain belonging to each of the different cereal taxa. In theory, a stacked bar chart might be the best format for the data, with four different portions in each bar corresponding to the four different cereals. In practice, however, it can be difficult to read comparative data in this format, so I have opted (as in McKerracher 2018) to plot the data for each cereal taxon on a separate graph, with the bars grouped by phase, and ordered by magnitude within each group to make the overriding trends more readily visible.

Table 20 – Relative proportions of taxa amongst grain in free-threshing grain-rich product samples.

sample	phase	nca	total grain	% barley	% wheat	% oat	% rye
ENB3	Early Saxon	Bedfordshire and Cambridgeshire Claylands	235	46.4	53.6	0.0	0.0
LE1	Early Saxon	Thames Valley	69	27.5	66.7	5.8	0.0
LLC10	Intermediate	Bedfordshire and Cambridgeshire Claylands	99	22.2	68.7	7.1	2.0
GAM2	Intermediate	Bedfordshire Greensand Ridge	219	15.1	80.4	1.4	3.2
IPS30	Intermediate	South Suffolk and North Essex Claylands	56	19.6	17.9	19.6	42.9
WKB1	Intermediate	South Suffolk and North Essex Claylands	211	4.3	91.5	4.3	0.0
Y22	Intermediate	Upper Thames Clay Vales	34	29.4	29.4	29.4	11.8
Y12	Intermediate	Upper Thames Clay Vales	43	55.8	37.2	7.0	0.0
LLC22	Mid Saxon	Bedfordshire and Cambridgeshire Claylands	57	22.8	54.4	19.3	3.5
SMB13	Mid Saxon	Breckland	558	0.4	0.0	0.7	98.9
SMB3	Mid Saxon	Breckland	39	25.6	7.7	5.1	61.5
HUT6	Mid Saxon	East Anglian Chalk	831	40.3	42.8	6.1	10.7
HUT3	Mid Saxon	East Anglian Chalk	611	26.8	67.8	2.8	2.6
HAM11	Mid Saxon	East Anglian Chalk	56	51.8	44.6	3.6	0.0
IPS3	Mid Saxon	South Suffolk and North Essex Claylands	226	11.1	35.8	4.4	48.7
IPS39	Mid Saxon	South Suffolk and North Essex Claylands	69	27.5	24.6	5.8	42.0
IPS26	Mid Saxon	South Suffolk and North Essex Claylands	74	13.5	41.9	2.7	41.9
IPS9	Mid Saxon	South Suffolk and North Essex Claylands	156	7.7	53.2	3.8	35.3
IPS2	Mid Saxon	South Suffolk and North Essex Claylands	254	6.7	46.5	13.0	33.9
IPS32	Mid Saxon	South Suffolk and North Essex Claylands	48	10.4	54.2	2.1	33.3
IPS4	Mid Saxon	South Suffolk and North Essex Claylands	44	61.4	6.8	6.8	25.0
IPS6	Mid Saxon	South Suffolk and North Essex Claylands	80	37.5	28.8	11.3	22.5

Sample	Phase	Region	Total	% 1	% 2	% 3	% 4
IPS11	Mid Saxon	South Suffolk and North Essex Claylands	152	42.1	21.1	18.4	18.4
IPS5	Mid Saxon	South Suffolk and North Essex Claylands	105	19.0	45.7	17.1	18.1
IPS12	Mid Saxon	South Suffolk and North Essex Claylands	45	33.3	40.0	13.3	13.3
IPS34	Mid Saxon	South Suffolk and North Essex Claylands	146	10.3	76.0	0.7	13.0
IPS22	Mid Saxon	South Suffolk and North Essex Claylands	72	11.1	65.3	13.9	9.7
IPS19	Mid Saxon	South Suffolk and North Essex Claylands	232	6.0	6.0	81.5	6.5
IPS21	Mid Saxon	South Suffolk and North Essex Claylands	309	51.5	40.8	6.8	1.0
LH4	Mid Saxon	Thames Valley	32	53.1	43.8	3.1	0.0
FOR1	Mid Saxon	Thames Valley	405	88.6	11.1	0.2	0.0
WFC14	Mid Saxon	The Fens	46	65.2	17.4	0.0	17.4
WFR1	Mid Saxon	The Fens	128	33.6	47.7	12.5	6.3
WFC1	Mid Saxon	The Fens	42	52.4	45.2	2.4	0.0
WLP7	Mid Saxon	The Fens	55	96.4	0.0	3.6	0.0
Y26	Mid Saxon	Upper Thames Clay Vales	80	35.0	46.3	7.5	11.3
Y49	Mid Saxon	Upper Thames Clay Vales	196	21.4	50.0	23.0	5.6
Y34-5	Mid Saxon	Upper Thames Clay Vales	424	33.3	52.6	10.1	4.0
Y30	Mid Saxon	Upper Thames Clay Vales	524	44.5	39.1	15.5	1.0
WST2	Generic	Breckland	294	0.0	0.0	0.0	100.0
HAM5	Generic	East Anglian Chalk	72	70.8	4.2	20.8	4.2
HAM33	Generic	East Anglian Chalk	608	77.8	18.8	2.1	1.3
HAM12	Generic	East Anglian Chalk	427	99.5	0.2	0.2	0.0
WRS4	Generic	Upper Thames Clay Vales	374	15.0	83.7	1.3	0.0

Table 21 – Chronological distribution of free-threshing grain-rich product samples.

phase	no. USG/FSP samples
Early Saxon	2
Intermediate	6
Mid Saxon	31
Generic	5

Table 22 – Geographical distribution of free-threshing grain-rich product samples.

National Character Area	no. USG/FSP samples
Bedfordshire and Cambridgeshire Claylands	3
Bedfordshire Greensand Ridge	1
Breckland	3
East Anglian Chalk	6
South Suffolk and North Essex Claylands	17
Thames Valley	3
The Fens	4
Upper Thames Clay Vales	7

The larger number of NCAs renders bar charts a less practical tool for the investigation of geographical patterns, which may be better served by a GIS approach. Since several sites are represented by more than one sample, a method is required which allows data for overlapping samples – i.e. those from the same site – to be equally visible. Inverse distance weighting (as in McKerracher 2018: 56) provides such a means of visualisation, interpolating geographical patterns from available data points (i.e. samples), taking into account the distances between them, and displaying the results as a shaded matrix or heat map (Chapman 2006: 76). Again, for ease of interpretation, the data are shown for each cereal taxon separately. It should be noted that the shading legend for each of these maps is different, to take account of the fact that the range of percentages is different for each cereal. For instance, because rye only rarely constitutes more than 30% of the grains in a sample, the maximum (black) end of the shading spectrum is set at 30%. This approach helps to bring out geographical trends within the relative productivity values of each different cereal taxon.

How well do the observations from the inclusive study stand up in light of the more exclusive results presented here? Taking chronological patterns first, the inclusive study reported that 'the proportions of wheat and barley grain do not change dramatically over time… More striking is the appearance of samples which are comparatively rich in rye and oat as a new phenomenon of the seventh century and later' (McKerracher 2018: 103–104).

Two of these patterns are closely repeated in the more exclusive study. The relative productivity of rye increases dramatically in the 8th and 9th centuries, while that of free-threshing wheat changes relatively little over time (Figures 30 and 31). Interestingly, a slightly different pattern emerges for barley: samples rich in barley appear in the main to be a Mid Saxon phenomenon (Figure 32). The pattern is not nearly so marked as that for rye, but it may still be said that the relative productivity of hulled barley appears to rise somewhat in the 8th and 9th centuries. For oat, meanwhile, an increase over time is less evident here than in the inclusive study (Figure 33). With the exception of one exceptionally oat-rich Mid Saxon sample (<IPS19> from Ipswich), it appears that oat maintains relatively low productivity throughout the period, with at most a very slight increase in the Intermediate and Mid Saxon phases.

What of the geographical patterns? The inclusive study (McKerracher 2018: 105) observed high concentrations of barley in the silt fens, Middle Thames valley, Suffolk coast, Breckland and East Anglian Heights, noting that these areas are characterised by salinity or the lighter, poorer soils tolerated by barley (cf. Murphy 2010: 215). This pattern is repeated in the exclusive study, excepting the Breckland and Suffolk coast (Figure 34), largely because the samples which contributed to these concentrations (from the Bloodmoor Hill and Redcastle Furze sites) were not eligible for this version of the analysis. Again, the inclusive study (McKerracher 2018: 105) observed concentrations of free-threshing wheat around the Bedfordshire and Cambridgeshire Claylands, the edges of the peat fens, the South Suffolk and North Essex Claylands, and the Severn, Avon and Upper Thames Clay Vales, noting how wheat is generally deemed well-suited to rich, clayey soils. The pattern is partly repeated here in the exclusive study, but without the westerly concentrations, largely because of the exclusion of two samples from Bishop's Cleeve (Glos) and Market Lavington (Wilts) respectively (Figure 35).

The 'scattered localised concentrations' of oats observed by the inclusive study (McKerracher 2018: 105), most noticeable around Yarnton and Ipswich, are also evident in the exclusive version (Figure 36). What is lacking in this version, however, is the marked concentration at Chadwell St Mary in south Essex (none of whose samples was eligible for this version of the analysis). And finally, the concentrations of rye around the 'sandy, droughty Breckland and the similarly dry and sandy Suffolk coast' highlighted by the inclusive study (McKerracher 2018: 105) are also clearly visible in this more exclusive version (Figure 37).

Apart from anything else, this comparison between the inclusive and exclusive studies highlights just how much regional and chronological patterns can hinge on the inclusion (or exclusion) of a relatively small number of sites and samples. This factor makes it all the more important for similar studies to be repeated in the future, when newly excavated sites and samples might substantially alter previously observed patterns.

The exclusive version of the analysis conducted here, although it has winnowed out some of the observations noted in *Farming Transformed*, can nonetheless draw similar overall conclusions. Above all, there still appears to be a genuine environmental gradient in the relative productivity of the different crops: drought-tolerant rye appearing most productive in the Breckland and near the sandy Suffolk coast; barley appearing most productive on relatively poor, well-drained or saline terrain; and wheat appearing most productive in the more clayey districts around the centre of the study regions. From a chronological perspective, there is little overall change in the relative productivity of wheat, which retains generally high productivity throughout the Early and Mid Saxon centuries. The relative productivity of oat remains low throughout the entire period, increasing only very slightly from the 7th and 8th centuries onwards. Barley shows an increase in relative productivity from around the 8th century onwards, showing similar values to wheat by the Mid Saxon phase. Rye's increase in relative productivity through the 7th and especially the 8th to 9th centuries is the most pronounced of all the diachronic patterns.

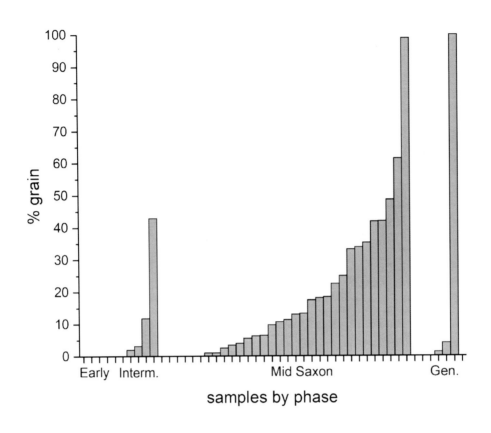

Figure 30 – Percentage of rye grains in free-threshing product samples, grouped chronologically.

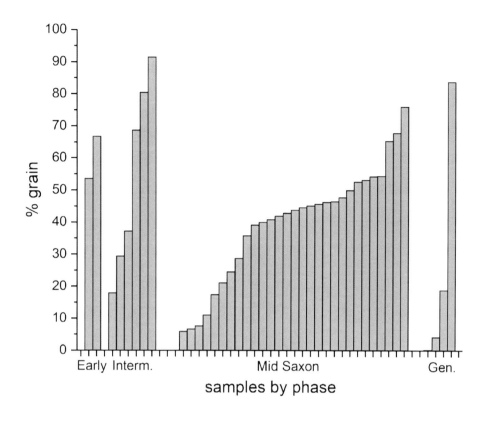

Figure 31 - Percentage of free-threshing wheat grains in free-threshing product samples, grouped chronologically.

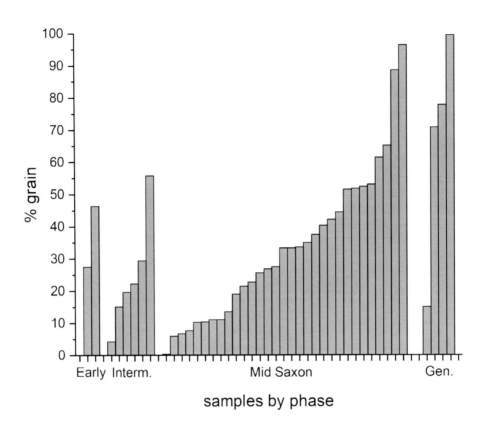

Figure 32 - Percentage of barley grains in free-threshing product samples, grouped chronologically.

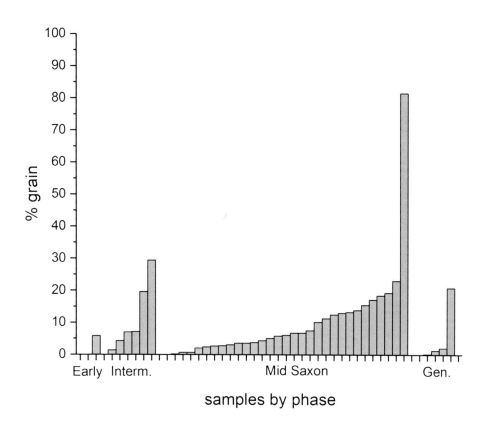

Figure 33 - Percentage of oat grains in free-threshing product samples, grouped chronologically.

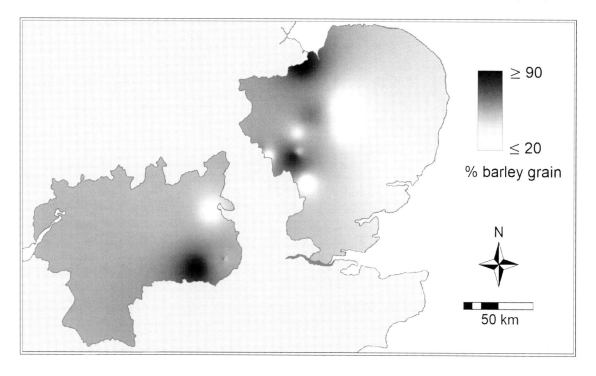

Figure 34 – Interpolated map of percentage of barley grains in free-threshing product samples.

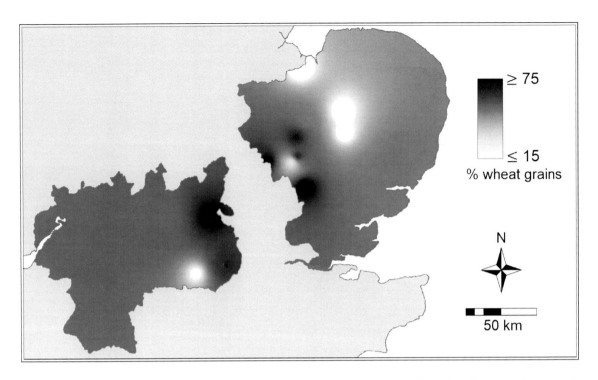

Figure 35 - Interpolated map of percentage of free-threshing wheat grains in free-threshing product samples.

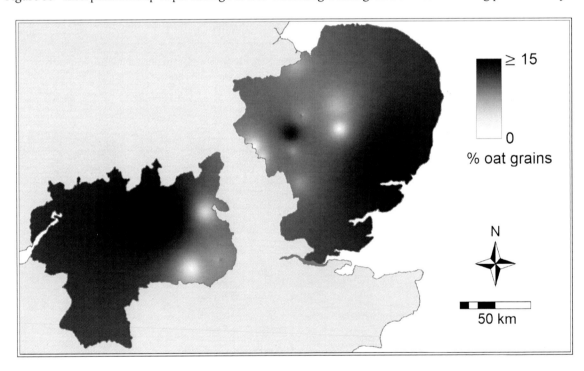

Figure 36 - Interpolated map of percentage of oat grains in free-threshing product samples.

Figure 37 - Interpolated map of percentage of rye grains in free-threshing product samples.

Relative productivity: glume wheats

Before discussing and contextualising these results in more detail, it is worth considering the relative productivity of glume wheats. As established in Chapter 4, the project dataset does not contain a sizeable group of samples which are both dominated by glume wheats and classifiable to a common crop processing stage. There is no glume wheat equivalent to the group of 44 'free-threshing grain-rich product' samples analysed in the previous section. Instead, we have three samples thoroughly dominated by glume wheats (two Early Saxon, one Mid Saxon), and 15 jointly dominated by glume wheats and free-threshing cereals (three Early Saxon, eight Mid Saxon, and four of Generic date). A variety of crop processing stages are represented among (and within) these samples.

Despite this taphonomic heterogeneity, however, some basic observations can be made regarding the proportions of different taxa in the cereal component of these samples. In the three samples dominated by glume wheats, although there are small amounts of oat, barley and rye present, spelt is the overwhelmingly dominant cereal (Table 23). It would be difficult to argue that anything but a high relative productivity for spelt is represented by these samples.

The relative productivity of crops in the 15 'mixed' samples is practically impossible to measure, since few if any of these samples can be persuasively assigned to single or even similar crop processing stages (as discussed in Chapter 4), and most are clearly dominated neither by a single plant part nor by a single taxon (Tables 24, 25 and 26). In the three samples from Lake End Road, for example, emmer is the best represented taxon among glume bases, rye tends to be the best represented among rachis segments (with not insignificant proportions of barley and free-threshing wheat), and barley tends to be the best represented among grains (with oat and rye reasonably well represented too). In the West Stow sample <WST1>, while the chaff component is dominated by spelt glume bases, the grain has similar proportions of barley, rye and wheat. Perhaps the nearest we can fairly come to observing a pattern is among the seven Mid Saxon and Generic grain-rich samples from Harston Mill, in which barley and spelt tend to be the best represented taxa, particularly barley, which is also the best represented taxon amongst the free-threshing 'product' samples from the same site. Such observations cannot, however, support any broader generalisations about the relative productivity of glume wheats over time and across the regions.

Table 23 - Relative proportions of cereal taxa, by glume bases and grains, in glume wheat samples.

taxa	AL1	BRT4	SMB1
% *Triticum dicoccum* Schübl./*spelta* L.	71.1	0.0	0.0
% *Triticum* L. indet.	0.0	33.3	44.3
% *Triticum spelta* L.	28.9	66.7	55.7
TOTAL GLUME BASES	**443**	**186**	**574**
% *Avena* L.	1.6	1.9	4.5
% *Hordeum* L.	7.1	24.5	72.7
% *Secale cereale* L.	0.8	0.0	0.0
% *Triticum dicoccum* Schübl./*spelta* L.	18.9	0.0	0.0
% *Triticum* L. indet.	29.9	73.6	22.7
% *Triticum spelta* L.	41.7	0.0	0.0
TOTAL GRAINS	**127**	**106**	**44**

Table 24 - Relative proportions of cereal taxa amongst glume bases in mixed cereal samples.

sample	total glume bases	% *Triticum dicoccum* Schübl.	% *Triticum dicoccum* Schübl /*spelta* L.	% *Triticum* L. indet.	% *Triticum* L. (free-threshing)	% *Triticum spelta* L.
CSE2	0	0.0	0.0	0.0	0.0	0.0
ENB2	4	50.0	0.0	50.0	0.0	0.0
HAM2	0	0.0	0.0	0.0	0.0	0.0
HAM26	0	0.0	0.0	0.0	0.0	0.0
HAM3	0	0.0	0.0	0.0	0.0	0.0
HAM32	0	0.0	0.0	0.0	0.0	0.0
HAM36	0	0.0	0.0	0.0	0.0	0.0
HAM4	0	0.0	0.0	0.0	0.0	0.0
HAM6-7	1	0.0	0.0	0.0	100.0	0.0
LW2-3	200	28.0	72.0	0.0	0.0	0.0
LW4	145	33.8	66.2	0.0	0.0	0.0
LW7	56	37.5	62.5	0.0	0.0	0.0
WST1	61	0.0	0.0	49.2	0.0	50.8
Y25	19	0.0	100.0	0.0	0.0	0.0
Y4	12	0.0	100.0	0.0	0.0	0.0

Table 25 - Relative proportions of cereal taxa amongst rachis segments in mixed cereal samples.

sample	total rachis segments	% *Hordeum* L.	% *Secale cereale* L.	% *Triticum dicoccum* Schübl.	% *Triticum* L. indet.	% *Triticum* L. (free-threshing)
CSE2	0	0.0	0.0	0.0	0.0	0.0
ENB2	6	0.0	33.3	0.0	0.0	66.7
HAM2	0	0.0	0.0	0.0	0.0	0.0
HAM26	0	0.0	0.0	0.0	0.0	0.0
HAM3	0	0.0	0.0	0.0	0.0	0.0
HAM32	0	0.0	0.0	0.0	0.0	0.0
HAM36	0	0.0	0.0	0.0	0.0	0.0
HAM4	0	0.0	0.0	0.0	0.0	0.0
HAM6-7	0	0.0	0.0	0.0	0.0	0.0
LW2-3	103	3.9	79.6	0.0	0.0	16.5
LW4	258	25.2	51.9	1.9	14.7	6.2
LW7	65	18.5	33.8	0.0	18.5	29.2
WST1	5	0.0	20.0	0.0	80.0	0.0
Y25	0	0.0	0.0	0.0	0.0	0.0
Y4	0	0.0	0.0	0.0	0.0	0.0

Table 26 - Relative proportions of cereal taxa amongst grains in mixed cereal samples.

sample	total grains	% Avena L.	% Avena sativa L.	% Hordeum L.	% Secale cereale L.	% Triticum dicoccum Schübl.	% Triticum dicoccum Schübl. /spelta L.	% Triticum L.	% Triticum L. (free-threshing)	% Triticum spelta L.
CSE2	42	2.4	0.0	35.7	0.0	0.0	31.0	0.0	31.0	0.0
ENB2	34	0.0	0.0	14.7	20.6	0.0	14.7	8.8	41.2	0.0
HAM2	560	1.3	0.0	45.4	0.7	0.0	0.0	21.3	0.4	31.1
HAM26	398	0.0	0.0	59.3	0.0	0.0	0.0	17.3	4.5	18.8
HAM3	46	10.9	0.0	50.0	0.0	0.0	0.0	13.0	4.3	21.7
HAM32	199	1.5	0.0	47.2	0.5	0.0	0.0	14.6	4.5	31.7
HAM36	177	0.0	0.0	33.9	0.0	0.0	0.0	19.2	1.1	45.8
HAM4	76	0.0	0.0	15.8	15.8	0.0	0.0	0.0	2.6	65.8
HAM6-7	195	28.2	0.0	32.8	0.5	0.0	0.0	2.6	5.1	30.8
LW2-3	203	9.4	0.0	39.9	22.2	2.5	7.9	12.3	5.9	0.0
LW4	411	8.0	0.0	59.4	1.9	5.1	15.3	5.6	4.6	0.0
LW7	221	34.4	0.5	31.2	0.9	7.7	3.6	10.9	10.9	0.0
WST1	75	6.7	0.0	28.0	32.0	0.0	0.0	33.3	0.0	0.0
Y25	290	1.0	0.0	74.8	0.0	24.1	0.0	0.0	0.0	0.0
Y4	29	3.4	0.0	72.4	0.0	0.0	3.4	17.2	3.4	0.0

Discussion

In this chapter I have attempted to gauge the relative importance of the different cereal crops in the dataset by parsing the concept of 'importance' into three properties, each accessible via a distinct archaeobotanical methodology: (i) prevalence, (ii) frequency of use, and (iii) relative productivity. The findings so far can be summarised as follows.

Hulled barley was highly prevalent – practically ubiquitous – throughout the 5th to 9th centuries, with no appreciable change over time. It was, however, used more frequently from the 7th century onwards, and its productivity increased notably in the 8th and 9th centuries, especially on the lighter, poorer soils upon which it can thrive better than wheat (Moffett 2006: 48, Table 4.3).

Wheat, like barley, was practically ubiquitous throughout the Early and Mid Saxon periods, and shows a similar increase in usage-frequency over time. Among the wheats, free-threshing varieties were practically as widespread as glume wheats in the Early Saxon period, each occurring at roughly 50% of sites. But whereas the distribution of spelt contracts between the Early and Mid Saxon periods, and that of emmer increases only slightly, free-threshing wheat becomes markedly more prevalent over time. Glume wheats were, on the whole, used less frequently from the 7th century onwards, while free-threshing wheats were used considerably more frequently from around this time. While spelt and emmer may have been particularly, locally productive at certain sites in both the Early and Mid Saxon periods, it is hard to gauge wider patterns on current evidence. Free-threshing wheat, meanwhile, maintained relatively high productivity throughout the 5th to 9th centuries, especially in clayey districts, upon whose heavier soils it is thought to thrive (Jones 1981: 107).

Oat is perhaps the hardiest of the cereals considered here. It was practically as prevalent as free-threshing wheat in the Early Saxon period, and became more widespread such that it retained parity (in prevalence) with free-threshing wheat in the Mid Saxon period. Over the same period, it came to be used more frequently. Oat may have been a relatively productive crop at Yarnton and Ipswich, but in general it increased only slightly in relative productivity over time.

Rye was not prevalent in the 5th and 6th centuries, but became much more widespread from around the 7th century onwards. It was used much more frequently from around the same time, and became much more productive in the 8th and 9th centuries, especially in sandy regions whose dry soils can be exploited by this deep-rooted, drought-hardy crop.

How do these observations compare with existing hypotheses about Early and Mid Saxon innovations in crop husbandry, as discussed in Chapter 1?

First, there is a question regarding the supplanting of glume wheats (principally spelt) by free-threshing wheat, a process normally thought to have been completed within the 5th century, but sometimes thought to have been completed only by the 8th century. The project dataset suggests that free-threshing and glume wheats were similarly widespread in the Early Saxon period, with free-threshing varieties only clearly becoming the most prevalent – and most frequently used – kind of wheat from the 7th century onwards. However, free-threshing wheat was always more productive than spelt (and glume wheats in general) throughout the 5th to 9th centuries.

Two interpretations of this pattern may be offered. The first posits that the relative productivity of spelt has been under-estimated, especially for the Early Saxon period, because there remains a widespread assumption that spelt does not 'belong' in post-Roman samples. Hence, all else being equal, and unless radiocarbon dates suggest otherwise, a sample rich in spelt grains will be assigned a Roman (or earlier) date. In this case, the rise of free-threshing wheat to become the more prevalent, productive and more frequently used of the wheats may have been only a gradual process conducted over the course of the 5th to 9th centuries.

Alternatively, if the low relative productivity of glume wheats is truly representative of the post-Roman archaeobotanical record, it would appear that although spelt remained relatively widespread in the Early Saxon period, it was contributing very little to most harvests in this period, and its prevalence was more likely due to its persistence as a very minor or volunteer crop. This minor or volunteer role declined over time, and the productivity of spelt was maintained at only a few exceptional localities, such as Thetford or Harston Mill.

What of the idea that free-threshing wheat supplanted hulled barley as the most important cereal crop from the 7th or 8th century onwards? The project dataset does indeed suggest that free-threshing wheat became more prevalent, and was used much more frequently, from the 7th century onwards. There is not, however, any corresponding contraction or reduction in the use of barley, which remains a ubiquitous and (increasingly) frequently used cereal crop throughout the Early and Mid Saxon periods. Moreover, there is no evidence to suggest that free-threshing wheats became a more productive component of harvests over this period, whereas hulled barley does appear to have become more productive, especially on lighter, poorer soils. These patterns would not suggest that crop husbandry strategies adapted to increase the productivity of wheat over barley during the 7th and 8th centuries, but rather vice versa, with barley yields improving through the exploitation of drier, poorer terrains.

Another theory is that oat and rye became more important crops from the 7th century onwards. The project dataset suggests that oat became more widespread and was used more frequently from the 7th and 8th centuries onwards, but did not become a much more productive component of cropping regimes during this period. One might then argue that it was being used more often, at more sites, but only as a minor crop – perhaps a hardy insurance crop. For rye, meanwhile, the project dataset shows considerable increases in prevalence, frequency of use, and relative productivity from the 7th and especially the 8th century onwards, particularly around the dry, sandy terrains of the Breckland and coastal Suffolk. Thus more often, and more widely, rye became an increasingly productive component of Anglo-Saxon cropping regimes, above all on the soils to which rye is well-suited but on which other cereals might struggle.

Finally, what of emmer, and its putative reintroduction as a local innovation of Saxon settlers in the Thames valley (Pelling and Robinson 2000)? The project dataset contains little more evidence than that presented by Pelling and Robinson. Emmer seems to have remained a rarely used and sparsely distributed crop, registering negligible increases in prevalence and usage-frequency over time, and appearing relatively productive only at Lake End Road and, perhaps, Yarnton. At Yarnton, the grains, glume bases and spikelets forks of emmer/spelt (emmer, where identifiable) were radiocarbon-dated to cal. AD 670–900. At Lake End Road, glume wheat grain and chaff – again mostly emmer, where identifiable – were radiocarbon-dated to cal. AD 435–663 in one instance, but the other samples were associated with eighth- to ninth-century pottery. It is perhaps plausible, therefore, to posit very localised emmer cultivation being practised at Yarnton and Lake End Road, potentially throughout the 5th to 9th centuries (thus agreeing with Pelling and Robinson), but elsewhere the evidence for emmer is so thin on the ground that it may well have occurred only as a rare volunteer or contaminant.

Summary

So far, this book has argued that the 7th and especially 8th centuries witnessed on the one hand a growth in surplus cereal production, and on the other hand regional changes in cropping choices, such that barley and above all rye became more often an increasingly productive component of the harvest, while oat was used more frequently as a minor crop. It would be reasonable to posit a connection between these trends, and argue that the changes in crop spectra were responsible for the increase in surpluses. However, to explore this idea further, we must consider the archaeobotanical missing link: cultivation environments and crop husbandry strategies, for which we rely on the witness of the arable weeds.

Chapter 6: The Witness of Weeds

Archaeologists sometimes draw a distinction between artefacts and ecofacts, i.e. the manmade and the natural. But arable fields, and by extension the charred remains which represent them in the archaeological record, fall into neither category – or into both. For while sets of crops are chosen directly by the farmers, sets of arable weeds spring up unbidden within the particular ecosystems which the farmers create through their crop husbandry strategies. Arable fields in their different forms and environments provide different ecological niches, and thus incubate different weed floras that thrive in – or at least competitively tolerate – such conditions. Thus, for example, species which are not easily discouraged by frequent soil disturbance, such as those which can regenerate from root fragments, enjoy a competitive advantage in deeply ploughed fields.

All of which means that the weeds represented in archaeobotanical samples provide us with a (potentially very sensitive) proxy for growing conditions in the Anglo-Saxon fields: a vegetative key to the vexed questions which were raised in Chapter 1, such as those concerning sowing times.

Weed ecological data

I have therefore undertaken weed ecological studies in order to make inferences about growing environments and, by extension, crop husbandry practices in the study regions during the Early and Mid Saxon periods. There are various possible approaches to the study of weed ecology. In this book, an autecological approach is adopted, in which the ecological tolerances/preferences of individual taxa, as determined through modern field observations, are examined. Autecology is here considered preferable to a phytosociological approach (based upon recognized plant communities), since phytosociology would not necessarily allow for substantial changes in weed spectra since antiquity (Bogaard 2004: 5–7; Charles *et al.* 1997: 1151–1152).

Ecological data pertaining to the wild taxa in the dataset (class C, as defined in Chapter 3) were collected from two main botanical compendia: PLANTATT (Hill *et al.* 2004; http://www.brc.ac.uk/resources - accessed January 2019) and Ecoflora (Fitter and Peat 1994; http://www.ecoflora.org.uk - accessed January 2019). Data from these sources were supplemented through the consultation of selected literature (Ellenberg 1988; Grime *et al.* 1988; Stace 2010; Streeter 2010).

The principal ecological variables collected were life history (perennation); flowering time (onset and duration); and Ellenberg indicator values – adapted for British flora in PLANTATT – for soil dampness (F), acidity (R), and fertility (N). The Ellenberg indicator values operate on a numerical scale describing a species' ecological preferences, as perceived in modern field observations. Ellenberg numbers are not ideal as an index of weed ecology, and might be criticised for suggesting too simplistic an association between species and environments. While this is a valid criticism, the values are nonetheless used here as an indicative, rather than definitive, guide to species' preferred or tolerated habitats. Above all, because these variables are available as open access data, they offer an independent (if imperfect) metadataset suitable for repeatable quantitative analysis.

It must be acknowledged that the use of these autecological data entails some caveats. Principally, because the data are based upon field observations of the species' respective occurrences in different environments, they are largely descriptive rather than explanatory, and do not differentiate between 'the many environmental variables to which individual species are responding' (Charles *et al.* 1997:

1152). Hence, they do not permit a detailed analysis of growing conditions. Functional autecology (the 'FIBS' approach) addresses this deficiency by examining functional traits of weeds, and is therefore methodologically more rigorous and informative. Original applications of this approach require the accumulation of primary data, which is beyond the scope of this project.

The functional attribute of flowering onset and duration, however, has been used as an index of sowing times. This approach follows the findings of Bogaard, Jones, Charles and Hodgson, who conducted a functional ecological analysis of flora of known origin from modern German field surveys, and found flowering onset and duration to be the most successful discriminating variables for distinguishing seasonality in sowing (Bogaard *et al.* 2001). The basic principle is that species which are able to flower and set seed after the disturbance of springtime ploughing – i.e. annuals whose flowering time is late or long, and perennials which can regenerate from vegetative fragments – are at a competitive advantage among spring-sown crops, whereas early- and short-flowering species may thrive undisturbed among autumn-sown crops, unless they are weeded out by hand (Table 27; Bogaard *et al.* 2001: 1173). Flowering variables for annuals were therefore used in a discriminant analysis, comparing samples from the project dataset with the German control data used in the original study, in order to investigate sowing times (I am grateful to Professor Bogaard for making the control data available to me).

Table 27 - Relationships between the flowering onset and duration of annual arable weeds and the sowing times of crops (after Bogaard *et al.* 2001 p.1175, Table 3).

class	flowering onset/duration	offers competitive advantage in
early/short	Jan.-Jun., 1-3 months	autumn-sown fields
late	Jul.-Dec., 1-5 months	spring-sown fields
long	Jan.-Jun. >5 months	spring-sown fields
intermediate	Apr.-Jun., 4-5 months	autumn- and spring-sown fields

Weed ecological analyses

The literature review in Chapter 1 has highlighted certain arable weed species which, according to some studies, may be indicative of agricultural innovations in the Mid Saxon period, such as stinking chamomile as an indicator of resumed cultivation of heavy clay soils, and henbane as an indicator of middening. The occurrence and significance of these perceived 'indicator species' can be investigated using the semi-quantitative approaches outlined above for the study of crop species. It is preferable, however, to take account of a wider range of species than those few which have been predefined as indicators. Inferences based upon a wider range of taxa are less likely to be skewed by chance occurrences or contamination; they also circumvent the caveat that individual species' ecological preferences might have changed since antiquity, since it is unlikely that several different species will all have changed in the same way (Jones 1992: 135-137).

The wild/weedy species recorded in the dataset constitute a wide array of variables, and multivariate statistical approaches were therefore appropriate. As already noted, discriminant analysis was used to investigate seasonality in sowing. In addition to this, the exploratory technique of correspondence analysis was employed. Correspondence analysis is an ordination technique that reduces many variables to two axes which account for the majority of variation in the data. In this case, samples are ordered by their compositional variations (i.e. their constituent species), while species are ordered by variations in their occurrence among samples. The two axes are used to produce scattergraphs, displaying the principal trends in the data. The origin (centre) of the graph represents typical, average composition; distance from the origin increases with a sample's (or species') deviation from the norm. Equally, the proximity between two samples or species is a measure of their similarity in composition or distribution (Bogaard 2004: 92–94). Correspondence analyses were conducted using Canoco for Windows 4.51 and scattergraphs were produced with CanoDraw for Windows 4.1 (ter Braak and Smilauer 2002).

The following criteria were applied in correspondence analyses, in order to exclude small samples whose contents are unlikely to be representative of their parent economies, and rare taxa whose occurrence may obscure patterns among the more common species. Samples each required a minimum abundance of ten weed seeds, and species each required a minimum presence of three samples (Appendix 1: Parameters 10-11). These criteria were applied recursively until a core of eligible samples and species was attained. The term 'species' is here used literally: taxa identified only to the level of family or genus were not included, as families and genera may encompass various species with different ecological profiles.

Both of these methods – the discriminant analysis of sowing times, and the exploratory correspondence analysis – were applied only to the subset of 44 samples identified as grain-rich free-threshing cereal products (FSP and USG classifications as defined in Chapter 4). The reason for this restriction is that crop processing exerts a biasing influence on the weed species preserved in a sample, and samples of mixed functional origin may contain spurious co-occurrences of taxa that did not, in fact, constitute an arable weed flora in antiquity. In short, the weed species represented in each of the free-threshing cereal product samples may be thought plausibly to belong together: belonging not only with each other, but also with the crops amongst which they have been preserved.

Sowing times

The classification of weed species in terms of flowering habit (onset and duration) may vary slightly according to which reference materials are consulted. It is therefore important to detail the species and classifications that are being used in a given analysis, as a piece of key metadata (Appendix 2: Metadata 5).

Of the 44 samples selected in Chapter 4 as free-threshing grain-rich products, only 21 contain at least ten seeds of classifiable weed taxa, i.e. those classifiable in terms of flowering onset and duration. Out of the total number of classifiable taxa present in each sample, I have calculated the proportions of those taxa classified as 'early/short', 'late', or 'long' (Table 28), and appended this to the modern control data for the discriminant analysis (Bogaard *et al.* 2001).

As with the discriminant analysis of weed seed types described in Chapter 4, the weed ecological discriminant analysis was conducted using IBM SPSS Statistics version 25, applying the 'leave-one-out' option for greater rigour (IBM Corporation 2017). The results are shown in Table 29. Of the 21 samples analysed, only two were classified as representing spring-sown fields: <WLP7> and <Y22>, both deemed to be USG samples in Chapter 4.

Table 28 - Relative proportions of early/short-, late-, and long-flowering species in the free-threshing product samples.

sample	proportion of early/short-flowering species	proportion of late-flowering species	proportion of long-flowering species	total no. species
ENB3	0.4	0.4	0.0	7
FOR1	0.8	0.2	0.0	5
GAM2	0.5	0.5	0.0	2
HUT3	0.3	0.5	0.0	4
HUT6	0.5	0.3	0.0	6
IPS11	0.3	0.4	0.0	7
IPS19	0.3	0.5	0.0	8
IPS2	0.4	0.2	0.0	5
IPS3	0.4	0.1	0.0	7
IPS34	0.4	0.3	0.0	7
IPS9	0.3	0.4	0.0	8
SMB13	0.3	0.4	0.1	7
SMB3	0.3	0.5	0.0	4
WFR1	0.3	0.3	0.0	7
WLP7	0	0.7	0.0	3
WST2	1.0	0.0	0.0	3
Y22	0	0.8	0.0	4
Y26	0.3	0.7	0.0	3
Y30	0.4	0.4	0.1	8
Y34-5	0.1	0.4	0.0	8
Y49	0.2	0.5	0.0	6

Table 29 - Sowing seasonality analysis of free-threshing product samples, by discriminant analysis of weed flowering onset and duration (discriminant function rounded to three decimal places).

sample	interpretation	Function 1
ENB3	autumn	6.179
FOR1	autumn	13.243
GAM2	autumn	6.857
HUT3	autumn	3.009
HUT6	autumn	7.840
IPS11	autumn	3.980
IPS19	autumn	3.009
IPS2	autumn	7.086
IPS3	autumn	7.863
IPS34	autumn	7.021
IPS9	autumn	3.746
SMB13	autumn	2.125
SMB3	autumn	3.009
WFR1	autumn	4.822
WLP7	spring	-1.821
WST2	autumn	17.500
Y22	spring	-2.312
Y26	autumn	3.309
Y30	autumn	4.047
Y34-5	autumn	1.822
Y49	autumn	1.727

There is a predicted bias towards autumn-sowing indicator species among crop processing products, as opposed to by-products (Bogaard *et al.* 2005: 507). Since these two product samples, <WLP7> and <Y22>, contradict this bias, their classification cannot be dismissed as an artefact of crop processing, and deserves closer attention. Are these two samples distinctive in any other way?

<WLP7> derives from a ditch context at Walpole St Andrew, a fenland site, and is dated to the 8th or 9th century. Its free-threshing cereal grain component is dominated by hulled barley (96.4%), with a very small proportion of oat (3.6%). It has an average density of 25.2 items per litre. Meanwhile, <Y22> derives from a gully context at Yarnton, in the Upper Thames valley, and is dated to the later 7th or early 8th century. Its free-threshing cereal grain component is jointly dominated by wheat, hulled

barley and oat (29.4% each), with a much smaller proportion of rye (11.8%). It has an average density of 3.9 items per litre.

Hence, these two supposedly spring-sown samples have little in common in terms of their composition or local environment. They both post-date the Early Saxon period, but then again only one of the samples included in the discriminant analysis, <ENB3>, is of Early Saxon date. The results do not therefore provide a strong basis for inferring chronological patterns, but we may at least say that they present no evidence for spring-sowing in the Early Saxon period, and do not indicate that autumn-sowing was a Mid Saxon innovation.

The limited evidence similarly inhibits our ability to seek local or regional patterns. Walpole St Andrew has only a Mid Saxon phase, so we cannot make chronological comparisons at this locality, and no other samples from the silt fens were eligible to be included in this analysis. The nearest local comparison is <WFR1>, an autumn-sown sample from Ely inland in the peat fens, which is dominated by wheat and (to a lesser extent) hulled barley. At the very most, therefore, one could say that a local tradition of spring-sown barley in the silt fens could be consistent with the little evidence currently available in that region. At Yarnton, meanwhile, the spring-sown <Y22> is the only classified sample from the site's Intermediate phase; the four other Yarnton samples in this analysis, classified as autumn-sown, belong to that site's Mid Saxon phase, and tend to have modestly higher proportions of wheat among their cereal grains. A comparison of five or six samples provides only very slender evidence, but this little evidence cannot exclude the possibility of a shift towards autumn-sowing and wheat cultivation from around the later 8th century at Yarnton.

As van der Veen notes, spring-sowing in Britain is potentially disadvantageous, giving a lower yield than autumn-sown crops, 'owing to a reduction in the total photosynthesis', although spring-sowing might be preferred if autumn-sowing is prohibited by a severely cold winter, waterlogging, and/or insufficient labour and a resultant need to spread tasks out over two sowing seasons (van der Veen 1992: 130). Could spring-sowing therefore have been adopted at Walpole St Andrew as a risk-buffering strategy on soils which were more prone to waterlogging, with repeated marine transgressions in the wetlands of the silt fens? This idea would be compatible with earlier observations that the salinity of this environment would have favoured the cultivation of barley, as a more salt-tolerant crop (Murphy 2010: 215).

How do these observations compare with the ideas outlined in Chapter 1? There I quoted a proposition by Banham: 'If changing from barley to wheat meant growing winter corn for the first time, improved drainage might be vital to prevent the young plants standing with their feet in water over the winter, even on soils which were not particularly wet in the spring and summer' (Banham 2010: 183). The slim new evidence presented in this chapter, combined with that in Chapter 5, might suggest a slightly different model. According to the results of this study, neither of these trends – the widespread cultivation (and increased relative productivity) of wheat, and autumn sowing regimes – are demonstrable innovations of the Mid Saxon period, although we should remember that the Early Saxon period is severely under-represented in this dataset, and that there is a general bias towards autumn-sowing indicator species. Nonetheless, the idea of a Mid Saxon spring-sown barley regime in the silt fens, influenced by the local environment, is arguably consistent with the available evidence, although that evidence is not particularly rich. The spring-sowing of mixed crops in Yarnton around the later 7th century, meanwhile, would theoretically make sense as a risk-mitigation strategy: reducing the risk of winter waterlogging, and spreading the risk of crop failure over several different species. Whether or not either of these strategies represents a Mid Saxon innovation, however, remains debatable.

Correspondence analysis

In accordance with the criteria set out above, the correspondence analysis included only those samples which are thought to represent free-threshing grain products (FSP/USG), and those weed seeds that are identified to species-level (including those such as *Bromus hordeaceus/secalinus*, where two or more closely related species have similar ecological profiles). Of these, the analysis required that samples contain at least ten seeds belonging to these species, and that species were present in at least three samples. These criteria were applied recursively (i.e. to take account of the fact that removing one sample might render a species inquorate, and vice versa, necessitating recalculation), resulting in 17 species and 21 samples being eligible for inclusion. These species and samples are listed along with important metadata in Tables 30 and 31 respectively. The samples in question are largely of Mid Saxon date (17 of 21 samples), with only two Intermediate and one Early Saxon sample included. This bias can be explained by the generally greater abundance of charred material in the Mid Saxon period (see Chapter 4). Hence there is unfortunately little opportunity in this analysis to explore diachronic trends in weed ecology.

In the first run of the correspondence analysis, <WST2> appeared as an outlier because of its abnormally high proportion of corncockle seeds (*Agrostemma githago* L., 96 out of 98 seeds; Figures 38 and 39). The presence of this sample thus obscured other patterns in the data, so it was then omitted in order that underlying patterns among the remaining samples might be explored. This second run produced a much clearer distribution of samples and species, with distinct gaps and clusters becoming apparent (Figures 40 and 41).

Table 30 - Species included in correspondence analysis, with selected metadata (* denotes perennials which can regenerate from vegetative fragments).

species	code	life history	flowering habit (annuals)	F	R	N
Agrostemma githago L.	agrgit	annual	early/short	5	6	5
Anthemis cotula L.	antcot	annual	late	5	6	6
Atriplex patula L./prostrata (Boucher ex. DC.)	atriplx	annual	intermediate	6	7	7
Bromus hordeaceus L./secalinus L.	brohor	annual	early/short	4	6	4
Chenopodium album L.	chealb	annual	late	5	7	7
Eleocharis palustris (L.) Roem. & Schult./uniglumis (Link) Schult.	eleoch	perennial *	n/a	9	7	4
Fallopia convolvulus (L.) Á. Löve	fallop	annual	late	4	7	5
Galium aparine L.	galapa	annual	intermediate	6	7	8
Lapsana communis L.	lapcom	annual	intermediate	4	7	7
Malva sylvestris L.	malva	perennial	n/a	4	8	7
Medicago lupulina L.	medlup	annual/perennial	intermediate	4	8	4
Plantago lanceolata L.	plalan	perennial	n/a	5	6	4
Polygonum aviculare L.	polavi	annual	late	5	6	7
Rumex acetosella L.	rumact	perennial *	n/a	5	4	3
Rumex crispus L.	rumcri	perennial *	n/a	6	7	7
Spergula arvensis L.	spearv	annual	early/short	4	5	5
Tripleurospermum inodorum (L.) Sch. Bip.	tripleu	annual/perennial	late	5	6	6

Table 31 - Samples included in correspondence analysis, with selected metadata.

sample	nca	phase	crop processing classification	sowing time classification
ENB3	Beds and Cambs Claylands	Early Saxon	USG	autumn
FOR1	Thames Valley	Mid Saxon	USG	autumn
GAM2	Beds Greensand Ridge	Intermediate	USG	autumn
HUT3	East Anglian Chalk	Mid Saxon	USG	autumn
HUT6	East Anglian Chalk	Mid Saxon	USG	autumn
IPS11	South Suffolk and North Essex Claylands	Mid Saxon	USG	autumn
IPS19	South Suffolk and North Essex Claylands	Mid Saxon	FSP	autumn
IPS2	South Suffolk and North Essex Claylands	Mid Saxon	FSP	autumn
IPS3	South Suffolk and North Essex Claylands	Mid Saxon	USG	autumn
IPS34	South Suffolk and North Essex Claylands	Mid Saxon	USG	autumn
IPS9	South Suffolk and North Essex Claylands	Mid Saxon	FSP	autumn
SMB13	Breckland	Mid Saxon	USG	autumn
SMB3	Breckland	Mid Saxon	USG	autumn
WFR1	The Fens	Mid Saxon	USG	autumn
WLP7	The Fens	Mid Saxon	USG	spring
WST2	Breckland	Generic	USG	autumn
Y22	Upper Thames Clay Vales	Intermediate	USG	spring
Y26	Upper Thames Clay Vales	Mid Saxon	USG	autumn
Y30	Upper Thames Clay Vales	Mid Saxon	USG	autumn
Y34-5	Upper Thames Clay Vales	Mid Saxon	USG	autumn
Y49	Upper Thames Clay Vales	Mid Saxon	FSP	autumn

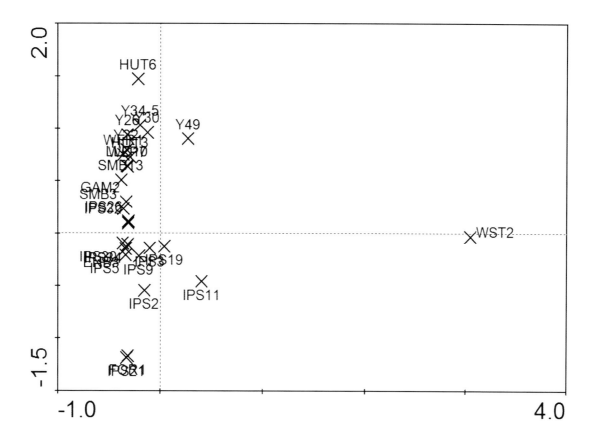

Figure 38 – Distribution of samples in correspondence analysis of free-threshing product samples and constituent weed species.

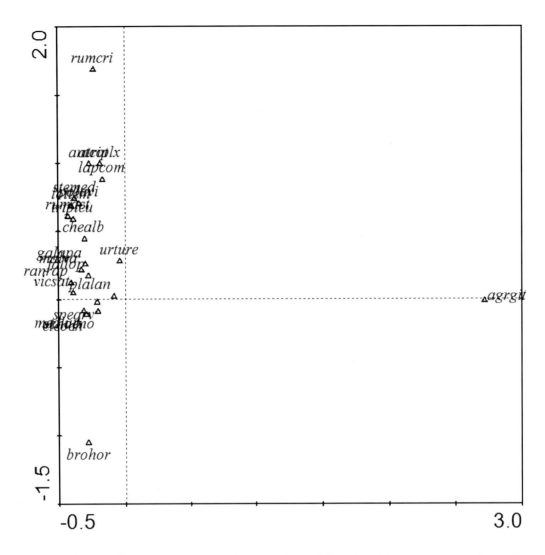

Figure 39 - Distribution of species in correspondence analysis of free-threshing product samples and constituent weed species.

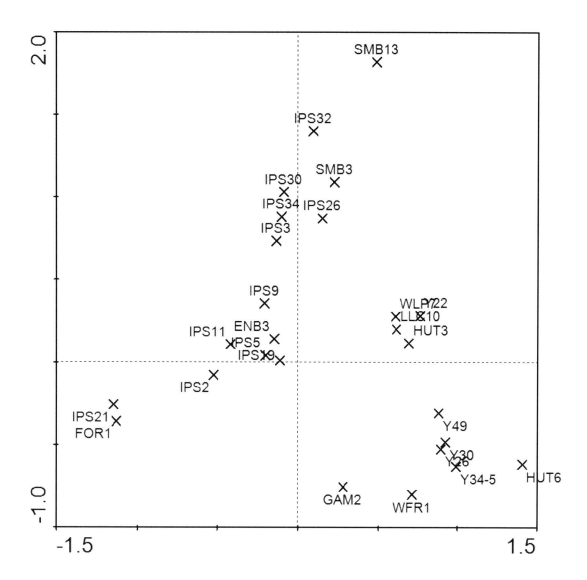

Figure 40 - Distribution of samples in correspondence analysis of free-threshing product samples and constituent weed species (excluding <WST2>).

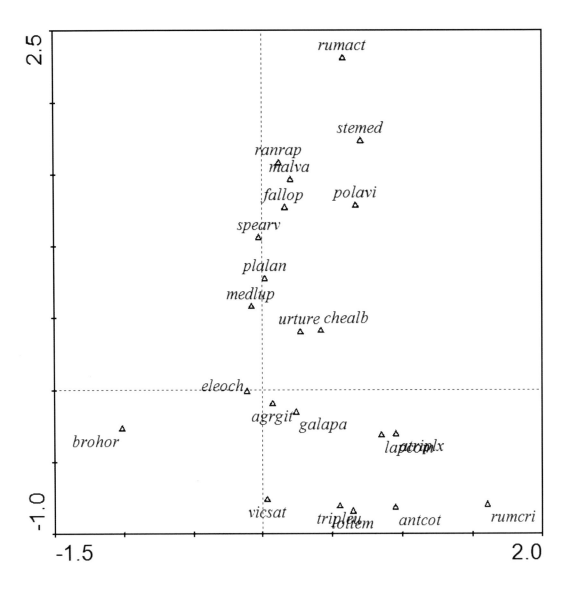

Figure 41 - Distribution of species in correspondence analysis of free-threshing product samples and constituent weed species (excluding <WST2>).

In the first instance, it is important to test whether any of the variation revealed in this analysis is due to differences in crop processing stages. In principle, this factor should not be decisive, since the analysis has been restricted to product samples. Nonetheless, it is theoretically possible that differences might exist between FSP and USG samples as a result of the additional sieving that the FSP samples have undergone. It is therefore important to exclude the possibility that the FSP and USG samples have been artificially separated on this basis. Figure 42 demonstrates that this is unlikely to be the case, since both USG and FSP samples are well-distributed throughout the graph.

If the aerodynamic properties of their weed seeds are not exerting a prime influence over the distribution of the samples, could the ecological profiles of the species be responsible? While any number of biological or ecological variables could be tested in this analysis, I have restricted my investigation to a set of readily available and easily quantifiable variables, as detailed in Table 30.

These include the flowering habit variable used in the discriminant analysis of sowing times. Although this is essentially the same variable, its usage here is somewhat different. The discriminant analysis was based on the presence or absence of particular weed species, to enhance comparability with the modern field surveys which provided the control data. In the correspondence analysis, however, quantities of seeds are taken into account. Neither approach is ideal: the presence/absence method could exaggerate the importance of a taxon represented by a single seed, while the quantified-seed approach could exaggerate the importance of a taxon which tends to produce more seeds. Investigation of flowering habits through correspondence analysis can therefore be considered complementary to the discriminant analysis above.

The discriminant analysis above classified only two samples as spring-sown. It is therefore surprising that the correspondence analysis appears to show a gradient of variation associated with flowering habit (Figure 43). Samples to the negative (left-hand) end of the x-axis contain higher proportions of seeds from species with early or short flowering seasons: that is, species most associated with autumn-sowing. Samples further to the positive (right-hand) end of the x-axis contain higher proportions of species with late or long flowering seasons, and perennials which can regenerate from vegetative fragments: that is, species most associated with spring-sowing. Those samples near the origin on the x-axis, and further to the positive (top) end of the y-axis, have more ambiguous contents in this regard. It is worth noting, too, that if we code samples by their sowing time classification from the discriminant analysis, the two samples there classified as spring-sown do form part of the 'spring-like' group at the positive end of the x-axis (Figure 44).

Perennation or life history – that is, whether a flower is an annual or perennial – has been cited as a proxy for soil disturbance in the study of Yarnton's plant remains, with an increase in annuals taken as a potential indicator of heavier ploughing (Stevens in Hey 2004: 363–364). However, flowering habit (as a proxy for sowing time) appears to be a more influential ecological variable than perennation in this case: there is no particular separation between samples rich in the seeds of annual species, and those rich in the seeds of perennial species, in this analysis (Figure 45).

What of the Ellenberg numbers? Moisture preferences (F numbers) perhaps account for some of the variation on the x-axis, with those samples at the negative (left-hand) end having higher proportions of 'drier' taxa and those at the positive (right-hand) end having more seeds of 'moister' taxa (Figure 46). However, the strongest separation is between F4 and F5/F6 taxa, and therefore any variability in moisture preferences could only be very slight. Nitrogen preferences (N numbers) seem to account for a greater degree of variation on the x-axis, with the less nitrophilous species better represented towards the negative end, and the more nitrophilous species better represented in the samples at the positive end (Figure 47).

It is harder to identify a compelling ecological gradient corresponding to the variation on the y-axis. There could perhaps be a weak trend relating to acidity: 'R' values appear to be slightly higher among the samples towards the positive (top) end of the y-axis, denoting greater tolerance of alkaline soils, and somewhat lower towards the negative (bottom) end, denoting greater tolerance of acidic soils. However, this pattern is neither very strong nor particularly consistent (Figure 48).

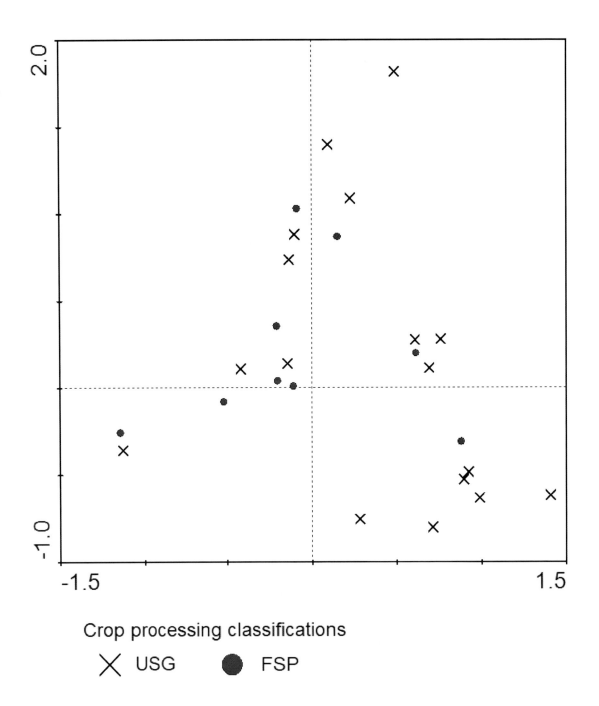

Figure 42 - Distribution of samples in correspondence analysis of free-threshing product samples and constituent weed species, coded by crop processing classification.

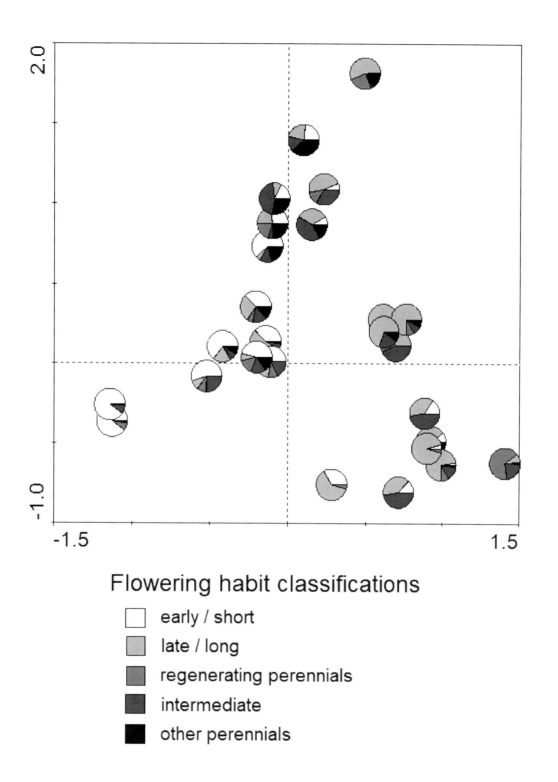

Figure 43 - Distribution of samples in correspondence analysis of free-threshing product samples and constituent weed species, coded by flowering habits of weeds.

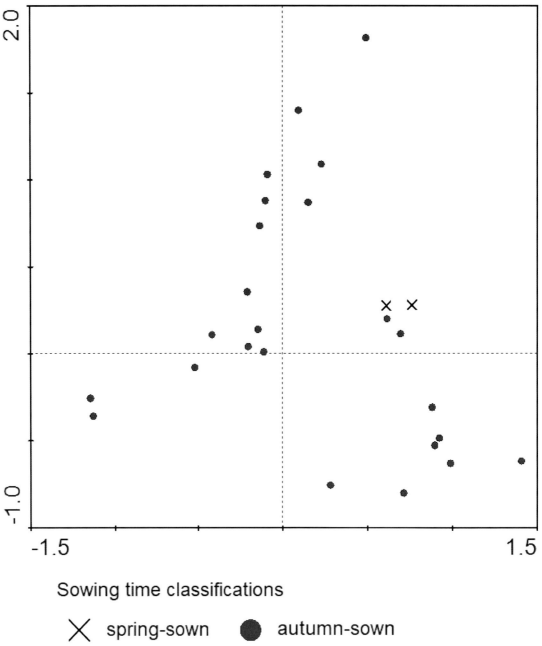

Figure 44 - Distribution of samples in correspondence analysis of free-threshing product samples and constituent weed species, coded by sowing time classification of samples.

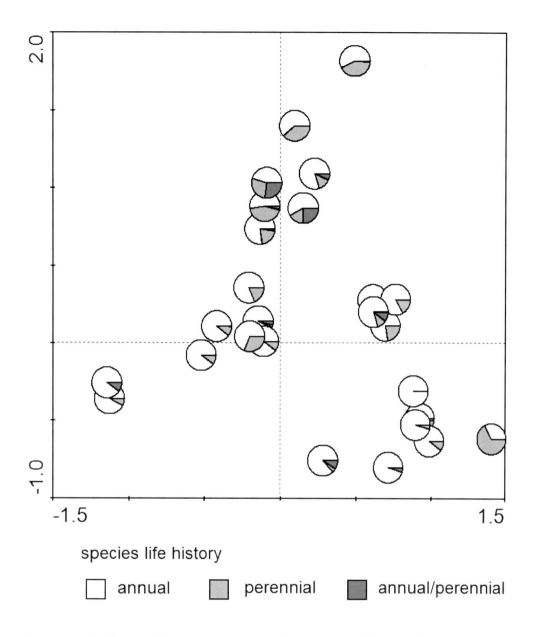

Figure 45 - Distribution of samples in correspondence analysis of free-threshing product samples and constituent weed species, coded by life history of weeds.

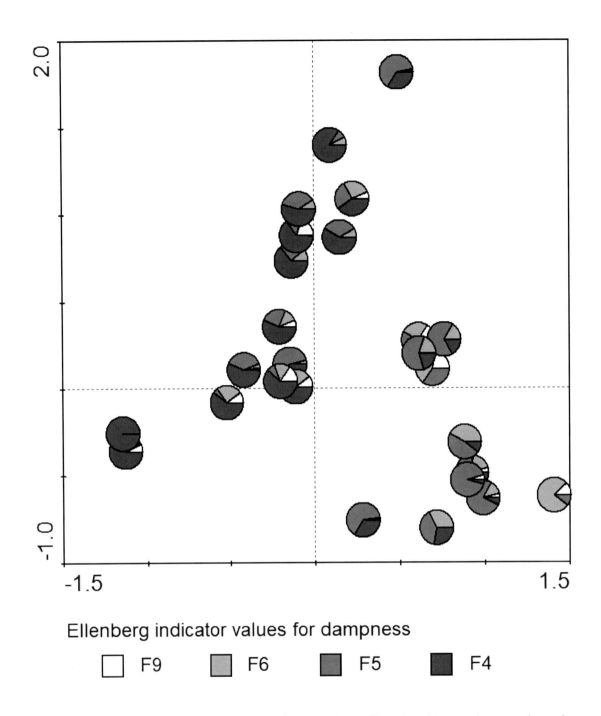

Figure 46 - Distribution of samples in correspondence analysis of free-threshing product samples and constituent weed species, coded by moisture preferences of weeds.

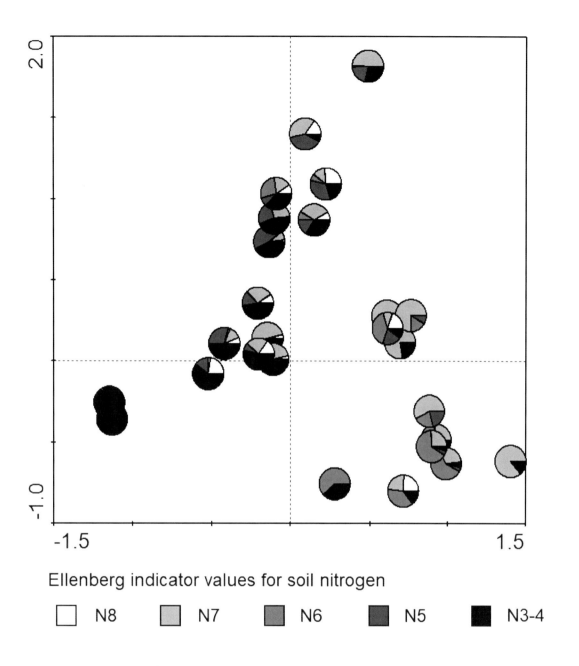

Figure 47 - Distribution of samples in correspondence analysis of free-threshing product samples and constituent weed species, coded by nitrogen preferences of weeds.

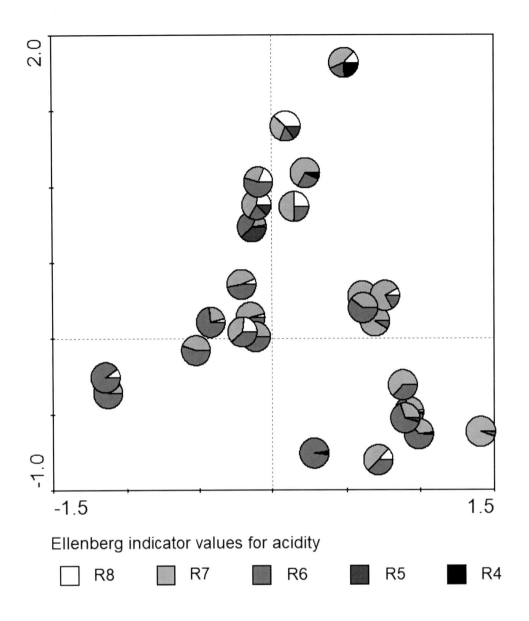

Figure 48 - Distribution of samples in correspondence analysis of free-threshing product samples and constituent weed species, coded by acidity preferences of weeds.

A more nuanced interpretation can be proposed for the seven samples drawn out towards the positive end of the y-axis: <IPS3>, <IPS26>, <IPS30>, <IPS32>, <IPS34> and <SMB3> and <SMB13>. These are labelled group A in Figure 49. Among the weed species most strongly associated with the positive end of the y-axis are corn spurrey (*Spergula arvensis* L.), wild radish (*Raphanus raphanistrum* L.) and sheep's sorrel (*Rumex acetosella* L.), species which are frequently associated with acidic sands and other non-calcareous soils (Clapham *et al.* 1962: 133–134, 257–258; Stace 2010: 418, 446, 467). The other species in this part of the graph, such as common mallow (*Malva sylvestris* L.), are more generally associated with disturbed and rough ground.

Two of the seven samples in this part of the graph are from Brandon, a Breckland site, and the other five are from Ipswich, whose immediate hinterland includes the sandy coast of Suffolk. Thus, it is possible to argue, with due caution given the small size of the available dataset, that these seven samples could represent sandier growing environments than those represented by samples closer to the origin and the negative end of the y-axis.

It is worth considering which other species contribute to the separation of other groups of samples. The two samples in group B, for instance, are chiefly separated from those of group C because their weed seeds are heavily dominated by *Bromus hordeaceus/secalinus* L. As noted in Chapter 3, brome seeds are common crop mimics, but could also potentially have been gathered deliberately – as fodder, for instance. The short dormancy of brome seeds has also led Jones to associate them with shallow cultivation (Jones 2009). Meanwhile, two of the most influential species separating group D from group E are *Anthemis cotula* L. and *Rumex crispus* L., the seeds of which are particularly well represented in the group D samples. *Anthemis cotula* is a species notably associated with heavy clay soils (Kay 1971: 625); and while *Rumex crispus* can thrive on a variety of soil types, there is some evidence for a propensity towards heavy soils (Cavers and Harper 1964: 758).

With this in mind, one might discern an ecological separation between these groups. The samples at the positive, right-hand end of the x-axis (groups D and E) have higher proportions of seeds from species associated with spring-sowing, slightly moister conditions, and soils richer in nitrogen. Among these samples, those in group D may be particularly strongly associated with heavier soils. Meanwhile, the samples at the negative, left-hand end of the x-axis (groups B and C) have higher proportions of seeds from species associated with autumn-sowing, slightly drier conditions, and relatively nitrogen-poor soils. Among these, the dominance of group B's samples by *Bromus hordeaceus/secalinus* could be due to a number of factors – such as its deliberate gathering, or particularly low soil disturbance – but is nonetheless consistent with the overall 'autumn-dry-poor' tendency.

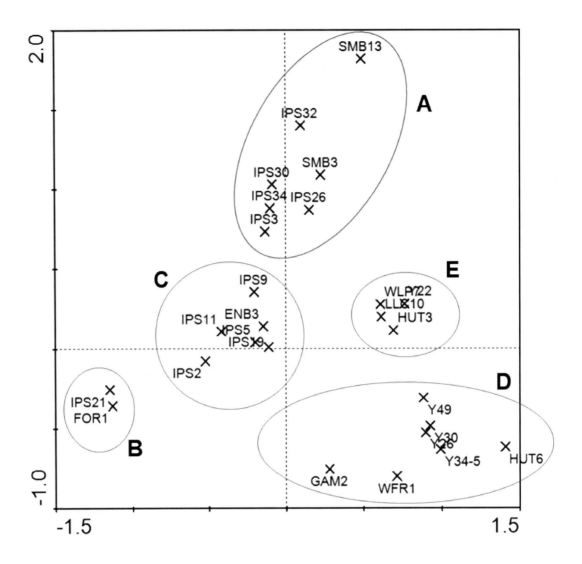

Figure 49 - Distribution of samples in correspondence analysis of free-threshing product samples and constituent weed species, divided into broad compositional groups.

What, if anything, can this ecological grouping exercise tell us about farming practices? It was argued in Chapter 5 that there is some correlation between the relative productivity of crops and the terrains on which they were grown: for instance, rye being most productive on drier, sandier soils. The possibility was thus raised that greater cereal surpluses were attained from the 7th and especially 8th century onwards by the matching of terrains to the crops which could grow most productively on them; or by the 'fine-tuning' of crop spectra to make most productive use of the land (McKerracher 2016).

The inferences in Chapter 5 were based, however, upon broad environmental profiles of sites and their hinterlands, whereas weed ecological data can shed more specific light on the arable environments from which individual samples derive. We can combine the correspondence analysis with the relative productivity data collated in Chapter 5, by coding the samples in the correspondence analysis scattergraph by the percentages of grain represented by a given cereal (see Chapter 5, Table

20). In this case, four versions of the correspondence analysis scattergraph have been produced, one each for barley, wheat, oat and rye. The size of the circle representing each sample corresponds to the percentage of grain in that sample belonging to the crop in question. Thus in Figure 50, each circle represents the percentage of rye grains in each sample, and it is clear that higher percentages of rye grains are found in those samples towards the positive (top) end of the y-axis: those group A samples which, as argued above, might well represent crops grown on acid sandy soils or other non-calcareous terrains. Hence, the weed data independently reinforce the idea that rye productivity was increased through the exploitation of sandy environments.

Beyond group A, crop-weed correlations are less clear. Oat shows very little patterning in this regard. The one sample which is heavily dominated by oat grains is that closest to the origin, i.e. closest to the norm (Figure 51). Similarly, there is no single pattern governing barley-rich samples, which occur in both groups B and E; samples in groups C and D also differ little in terms of their barley-richness (Figure 52). Wheat-rich samples are similarly heterogeneous in their distribution: similar proportions of wheat grain are found in samples across all groups, but the samples with the most consistently high percentages of wheat are those in groups D and E (Figure 53).

These observations give rise to some important points. First, if we interpret the flowering habit gradient as a proxy for sowing seasonality, then there is no evidence for wheat and rye being autumn-sown crops and barley and oat being spring-sown crops, as might be predicted from later medieval crop rotation traditions. Rather, there is a very slight tendency for wheat to appear most productive amongst the putative spring-sown samples at the positive end of the x-axis. This group of samples might also be characterised by heavier, richer, perhaps slightly moister soils, so that the association between this group and relatively high wheat productivity is consistent with the predicted propensity for wheat to thrive on heavy soils.

The evidence would thus be compatible with a scenario in which heavier soils were being exploited for wheat cultivation, but could not always be kept sufficiently well-drained to allow wheat to be sown in the autumn and remain safe from winter waterlogging. Autumn-sowing was perhaps favoured, however, on the lighter, poorer terrains represented at the negative, left-hand end of the x-axis, where wheat, barley, oat and rye might all have been autumn-sown and (comparatively) safe from winter waterlogging because of the better natural drainage. One implication of this might be that artificially improved drainage of heavy soils – by deep ploughing and the creation of ridge and furrow – was not commonly (or effectively) practised at this time.

While it must be stressed that the above interpretations can be made only tentatively, because of the relatively small subset of data being utilised, it is nonetheless possible to propose a meaningful model to explain the emergent patterns. Three main clusters of samples have been identified. First, there is an autumn-sown cluster, particularly associated with Ipswich, characterised by low nitrogen preferences and heterogeneous crop composition. Second, there is a spring-sown group – associated with sites such as Yarnton in the Upper Thames Clay Vales and Ely in the peat fens – characterised by higher nitrogen preferences, slightly higher moisture preferences, and some tendency towards wheat-richness. *Anthemis cotula* and *Rumex crispus* may indicate the use of heavy clay soils among most samples in this group. Finally, there is a sandy group of ambiguous seasonality – associated with Ipswich and Brandon – with a tendency towards relatively high proportions of rye. It is notable that Ipswich is well represented in both the sandy group and the autumn-sown group. It may well be, therefore, that Ipswich was in receipt of cereals from at least two different parts of its hinterland: the sandy Suffolk coast, and the slower-draining inland plains, whether simultaneously or successively within the Mid Saxon period.

In this context, spring-sowing could be seen as part of a risk-buffering strategy in environments prone to harsh winters and/or waterlogging. By contrast, in the supposed autumn-sown group of samples, the varied spectrum of cereals could itself represent an alternative risk-buffering strategy on lighter soils, whether directly through maslin cultivation (i.e. growing several crops together at once), or less directly through the centralised accumulation of separately cultivated monocrops of wheat, barley, rye and oat. The adoption of risk-buffering strategies such as these would be consistent with a postulated arable expansion in the Mid Saxon period, with local crop husbandry strategies being adapted to suit more challenging terrains as they were brought under the plough, both in the drier regions of East Anglia and on the wetter ground of the Upper Thames Clay Vales.

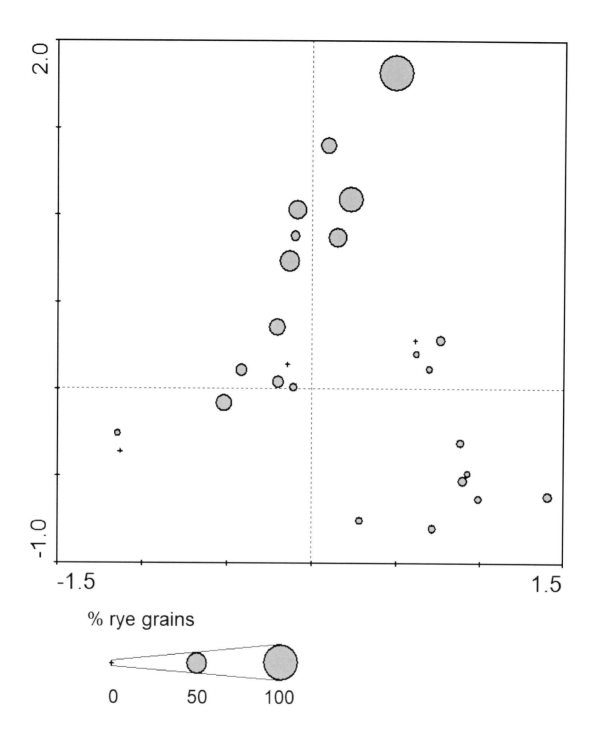

Figure 50 - Distribution of samples in correspondence analysis of free-threshing product samples and constituent weed species, coded by percentage of rye grains amongst grain content.

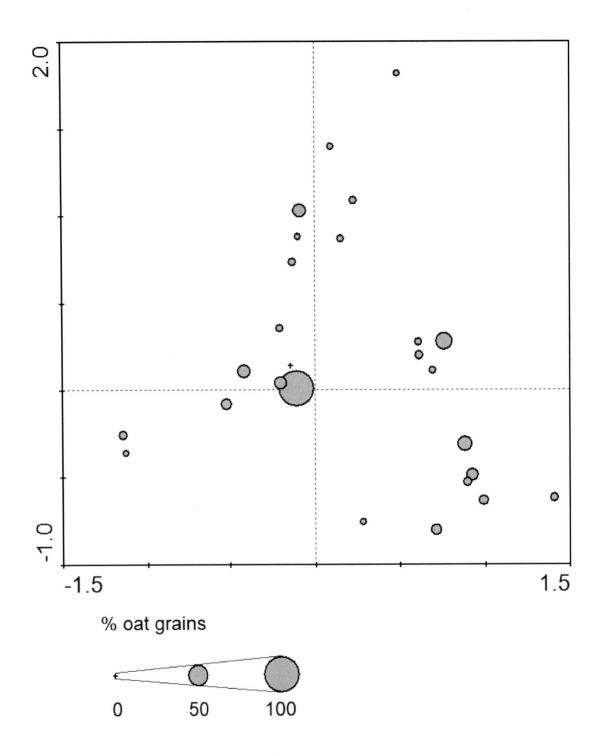

Figure 51 - Distribution of samples in correspondence analysis of free-threshing product samples and constituent weed species, coded by percentage of oat grains amongst grain content.

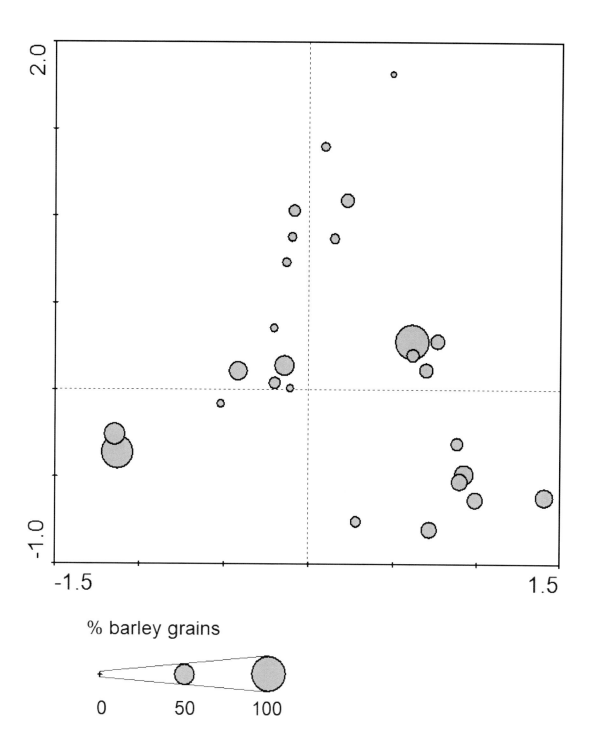

Figure 52 – Distribution of samples in correspondence analysis of free-threshing product samples and constituent weed species, coded by percentage of barley grains amongst grain content.

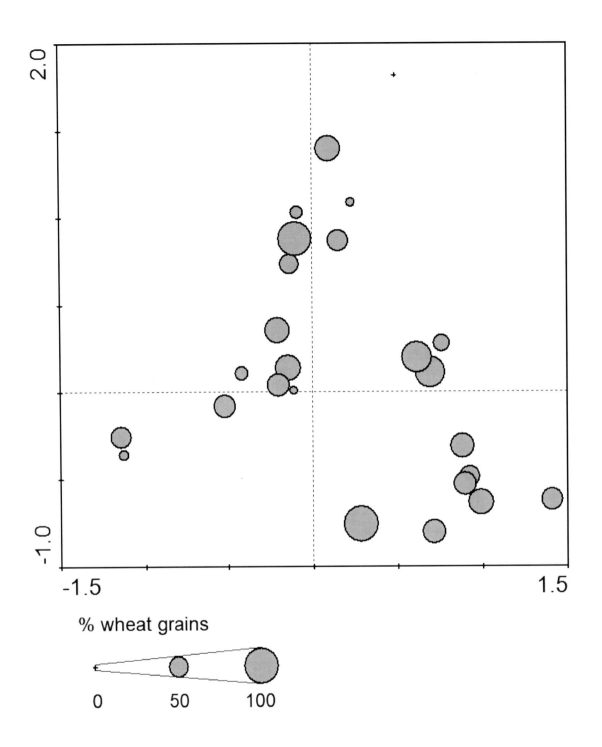

Figure 53 - Distribution of samples in correspondence analysis of free-threshing product samples and constituent weed species, coded by percentage of free-threshing wheat grains amongst grain content.

Changes in presence

Useful as the multivariate methods employed above can be, they have a limited scope, shedding light on only a small proportion of the project dataset. The net can be cast much wider, however, via a presence analysis of weedy taxa. Presence analysis is a technique used in ecological surveys to gauge the prevalence of a species within a certain environment, such as an arable field. The method entails the calculation of the proportion of sample units – such as randomly allocated grid squares – within which each species occurs. It may reasonably be inferred that a species which occurs in a high proportion of sample units enjoys a competitive advantage in that environment. When this method is applied to archaeobotanical material, the units in question may be soil samples or site assemblages. In neither case is the analogy with ecological surveys particularly close, because there is no reliable means of studying biological populations in the way that modern field surveys do. Even where, as in Chapter 4 and in the multivariate statistical analyses above, we can attempt to select samples which may conceivably each represent a single harvest, collectively these still represent different arable fields spread across time and space. Hence, the only 'population' that we can study is the time-averaged, multi-generational, multi-site population of the arable fields from a period.

Presence analyses were employed in Chapter 5 in the context of gauging the relative importance of different crop species. In that case, presence among assemblages was taken as a proxy for a crop's prevalence, while presence among samples was taken as a proxy for a crop's frequency of use. While it is still meaningful to speak of the prevalence of a weed in a given period, it hardly makes sense to speak of a weed's frequency of use. Rather, frequency of contamination might be a better phrase or concept.

In terms of both assemblages and samples, the presence values of most weedy taxa increase somewhat between the Early and Mid Saxon periods. To some extent, this pattern may simply reflect the generally greater abundance of charred plant remains that survive from the latter period: the Mid Saxon dataset is larger, and therefore the probability of any taxon's occurrence is, in theory, correspondingly greater. The principal interest here is in those species whose presence values change to an unusual extent over time. Only species-level identifications (including those embracing two or three closely-related species) are considered, to circumvent the problematising effects of overlapping categories at genus- and family-level. After examining the range of values for increases or decreases in presence between the Early and Mid Saxon periods, it was judged that a change in presence of at least ten percent was a practical quorum for defining significant changes (Appendix 1: Parameter 12).

The results of these presence analyses are shown in Table 32 for prevalence and Table 33 for frequency of contamination. Most of these species increase progressively in both prevalence and frequency of contamination through the Early Saxon, Intermediate and Mid Saxon phases, although a few (e.g. corncockle, *Agrostemma githago*) undergo some contraction in the Intermediate phase before a steep Mid Saxon rise. While the listed species do not seem to constitute a single, coherent group in ecological terms, certain trends are apparent. Several of the species with the most conspicuous rises in prevalence are relatively nitrophilous (Ellenberg N7-9): common or spear-leaved orache (*Atriplex patula/prostrata*), common mallow (*Malva sylvestris*), fat hen (*Chenopodium album*), knotgrass (*Polygonum aviculare*), black nightshade (*Solanum nigrum*) and especially stinging nettle (*Urtica dioica*) and henbane (*Hyoscyamus niger*). Spike-rush (*Eleocharis palustris/uniglumis*) indicates wet soils, and stinking chamomile (*Anthemis cotula*) is characteristic of heavy clay soils (Kay 1971: 625).

Overall, therefore, these results suggest that from the 7th and especially 8th century onwards, heavy clay soils, wet soils, and nitrogen-enriched soils become better represented in the ecological profiles of charred plant remains. Over this period, species which grow well on such soils became more prevalent and contaminated harvests more frequently. While not directly indicative of any specific innovations, these patterns would be consistent with the extension of arable onto damp, previously uncultivated clays and an intensification of soil-enrichment strategies, such as manuring, middening, or folding, to keep arable land 'in good heart' – though not necessarily at the same time or in the same place (Williamson 2003: 79–81).

Table 32 - Assemblage-based presence analysis of weed species, showing those with at least 10% change between the Early and Mid Saxon periods.

species	Early-Mid Saxon % change	% presence (assemblages)			
		Early Saxon	Intermediate	Mid Saxon	Generic
Atriplex patula L./*prostrata* (Boucher ex. DC.)	28.4	21.6	34.8	50.0	23.1
Malva sylvestris L.	23.3	13.5	30.4	36.8	7.7
Anthemis cotula L.	23.0	24.3	47.8	47.4	15.4
Agrostemma githago L.	20.8	10.8	4.3	31.6	23.1
Eleocharis palustris (L.) Roem. & Schult./*uniglumis* (Link) Schult.	20.3	27.0	43.5	47.4	7.7
Chenopodium album L.	20.3	29.7	34.8	50.0	38.5
Polygonum aviculare L.	20.3	29.7	21.7	50.0	7.7
Raphanus raphanistrum L.	18.3	5.4	13.0	23.7	0.0
Odontites vernus (Bellardi) Dumort.	15.6	5.4	4.3	21.1	7.7
Urtica dioica L	13.2	0.0	8.7	13.2	0.0
Solanum nigrum L.	13.2	0.0	4.3	13.2	7.7
Hyoscyamus niger L.	13.0	5.4	0.0	18.4	0.0
Ranunculus repens L.	10.5	0.0	4.3	10.5	15.4
Reseda luteola L.	10.5	0.0	0.0	10.5	0.0
Lapsana communis L.	10.5	0.0	0.0	10.5	0.0
Plantago major L.	10.5	2.7	4.3	13.2	0.0
Rumex crispus L.	10.4	5.4	8.7	15.8	0.0
Spergula arvensis L.	10.4	5.4	8.7	15.8	7.7
Medicago lupulina L.	10.4	5.4	4.3	15.8	7.7
Plantago lanceolate L.	10.0	18.9	30.4	28.9	23.1
Total number of assemblages (n=111)		**37**	**23**	**38**	**13**

Table 33 - Sample-based presence analysis of weed species, showing those with at least 10% change between the Early and Mid Saxon periods.

species	Early-Mid Saxon % change	% presence (samples)			
		Early Saxon	Intermediate	Mid Saxon	Generic
Anthemis cotula L.	14.6	5.3	15.9	19.9	7.0
Atriplex patula L./*prostrata* (Boucher ex. DC.)	13.2	4.8	8.6	18.1	5.3
Eleocharis palustris (L.) Roem. & Schult./*uniglumis* (Link) Schult.	12.2	8.7	12.6	20.9	1.8
Chenopodium album L.	12.0	9.2	12.6	21.2	15.8
Bromus hordeaceus L./*secalinus* L.	10.0	12.1	13.9	22.1	22.8
Total number of samples (n=736)		207	151	321	57

Summary

This chapter has taken three complementary approaches to archaeobotanical weed ecology, as a proxy for crop husbandry strategies in the Early and Mid Saxon periods: discriminant analysis to investigate sowing times; exploratory correspondence analysis to investigate unpredicted axes of variation; and presence analysis to investigate which weeds became more prevalent or frequent contaminants over time. The presence analyses suggest a general expansion of arable onto heavier, damper soils through the 7th and 8th centuries, together with an increase in nitrogen-richness. The multivariate statistical analyses add detail to this general picture, by suggesting that distinct groups of archaeobotanical samples may represent distinct crop husbandry strategies pursued in the Mid Saxon period: specialised rye cultivation in the sandy Breckland and Suffolk Coast; wheat-oriented spring-sown crops on heavier, damper soils; and the autumn sowing of a broader crop spectrum on lighter soils, especially in the hinterland of Ipswich, which was apparently receiving cereals from at least two different directions. With these models in mind, it is time to reconsider the hypotheses and research questions outlined at the start of this study.

Chapter 7: More than the Sum of their Parts

In the opening chapter, I proposed that the objective of this book is to establish how, when and where crop surpluses grew, crop spectra shifted, and crop husbandry regimes changed in Early and Mid Saxon England, through a set of related archaeobotanical investigations. As a starting point, I considered a number of current hypotheses in Anglo-Saxon archaeology, regarding how crop husbandry changed between the 5th and 9th centuries, focusing specifically upon those ideas that can be explored archaeobotanically. These I described in terms of four kinds of innovation:

1. **Crop biological innovations:** bread wheat supplanting spelt as the main wheat crop; bread wheat supplanting barley as the main cereal crop overall; emmer appearing as a localised innovation, and further diversification through the increased importance of rye and oat.
2. **Chemical innovations:** expansion onto heavier, more fertile soils; an increase in manuring or middening to maintain soil fertility.
3. **Mechanical innovations:** the increased uptake of heavy mouldboard ploughing.
4. **Managerial innovations:** the handling of greater crop surpluses; the advent of autumn sowing regimes.

The foregoing analyses have shed new light on these hypotheses, at least for the two case study regions upon which this book has focused.

Using the different aspects of 'importance' defined in Chapter 4, we can say that free-threshing wheat appears to have been much more productive than spelt as early as the 5th and 6th centuries. However, in terms of prevalence and frequency of use, free-threshing wheats only begin clearly to outstrip glume wheats (primarily spelt) from the 7th and 8th centuries onwards. Whether spelt persisted as a volunteer crop or a minor crop – or indeed occurs frequently as an archaeobotanical contaminant, spuriously appearing in Anglo-Saxon contexts – remains unclear, but the evidence from Harston Mill in particular suggests that we should not dismiss the idea of Mid Saxon spelt cultivation too readily. It remains possible that glume wheat cultivation lasted into the 8th century and beyond thanks to its naturally good preservation qualities, which may have been particularly desirable for the gathering and storing food renders at central places (see e.g. McKerracher 2017).

In no respect does free-threshing wheat outstrip hulled barley in importance in the Mid Saxon period. Even though free-threshing wheat increases in prevalence over time, hulled barley remains the most prevalent crop throughout the period. Free-threshing wheat, and wheats in general, were used more frequently from the 7th century onwards, but not necessarily any more frequently than barley by the 8th and 9th centuries. The relative productivity of free-threshing wheat does not increase appreciably over time but, perhaps surprisingly, hulled barley does seem to become more productive. These results are not necessarily incompatible with the documentary evidence adduced by Banham in support of the 'bread wheat thesis'. For example, food rents attest directly to the desirability of wheaten bread among landowners from around the 9th century onwards, whereas barley bread, according to Felix's eighth-century *Life of St Guthlac*, appears to have been appropriate sustenance for an ascetic anchorite, second only to starvation (Banham 2010: 176–177). The former evidence falls at, or beyond, the end of the Mid Saxon period; but even if an élite preference for wheaten bread, and a very low cultural value for barley bread, were current prior to the 9th century, these cultural values would not necessarily be manifested economically in wheat-dominated cropping regimes, let alone wheat-dominated charred crop deposits. The presumed low status of barley bread would not be inconsistent with its widespread consumption by lower-status consumers. Alternatively, or additionally, barley may have been – as Banham indeed suggests – the favoured crop for the brewing of beer, and therefore necessarily ubiquitous, regardless of the cultural status of barley bread (Banham 2004: 25–26).

Beyond wheat and barley, some studies have posited a trend towards crop diversification in the Mid Saxon period, with emmer, oat and rye among the crops that contributed to this trend. This study has found little additional evidence to augment that which underlies the argument of Pelling and Robinson (2000) that emmer may have become important as a local innovation in the Early and Mid Saxon periods. While the prevalence and usage-frequency of emmer are shown to increase through the 7th and 8th centuries, both values are persistently low, and thus consistent with the model of a localised rather than a common innovation.

Increases in prevalence and usage-frequency through the 7th and 8th centuries are much more pronounced for oats and especially rye. It would be fair to suggest that oat and rye were both common crops by the Mid Saxon period, albeit with some regional patterning (see Chapter 5: Figures 28 and 29 respectively). However, while rye also demonstrates a corresponding increase in relative productivity over this same period, this is less marked for oats. Thus, it might be argued that oat became more common as a minor crop – perhaps an insurance crop, given its hardiness.

It will be clear from the evidence presented in Chapter 5 that the free-threshing cereal samples tend to include grains from at least two different crop species, and this raises the question as to whether they represent the products of mixed cropping. Mixed cropping – variously known as maslin, mixtil, dredge or other terms depending upon the crops involved – cannot be conclusively demonstrated in archaeobotanical samples, but neither can the possibility be entirely excluded, especially when the majority of samples in this project's dataset show some degree of mixed cereal composition (Jones and Halstead 1995: 104; Slicher van Bath 1963: 263–264). Since one of the purposes of maslin cultivation is risk-buffering, through the juxtaposition of crops with different tolerances or preferences, one could conjecture that the perceived rise in the cultivation of oat and rye, both of which are tolerant of relatively poor growing conditions (Moffett 2006: 48, Table 4.3), may have served this purpose in Mid Saxon crop husbandry.

There are also geographical patterns among these data, which touch upon a chemical and managerial innovation recounted above: to wit, an expansion of arable land, particularly onto heavier, more fertile soils. With regard to free-threshing wheat, there is some tendency for relative productivity to be highest on heavier soils, but this does not in itself demonstrate that such soils were more frequently cultivated over time. Indeed, there are comparable indications that barley and rye achieved higher relative productivity on terrains to which they are ecologically well-suited.

It may be therefore that arable expansion embraced a range of different terrains – whether heavy, light, damp or dry – each of which was exploited in such a way as to obtain the most reliable yield. This process not only entailed localised modifications to crop spectra but also, as the weed ecological analyses have suggested, modifications to husbandry practices. Hence it could be that, contrary to popular wisdom, free-threshing wheat grown on heavier soils was sometimes spring-sown in order to reduce the risk of winter waterlogging, in a scenario where adequate drainage was not easily achieved. It is worth noting in this context that, in an eighth-century hagiographical anecdote recorded by Bede, it is implied that St Cuthbert's spring-sown wheat crop was sown at the proper time – *tempore congruo* – at least for his exposed hermitage off the coast of Northumbria (HE IV.28; VCB 19); while Comeau (2019) discusses evidence for a spring-sown wheat tradition in south Wales.

Finally, although the archaeobotanical evidence provides no direct evidence for heavy ploughing or middening or manuring, the increased prevalence of nitrophilous henbane and stinging nettle in the Mid Saxon period could indicate some effective means of soil fertilisation.

Our working hypothesis, then, is that agricultural development in Early to Mid Saxon England involved a tripartite diversification: cereal crop choices, arable terrains, and husbandry strategies may all have diversified in tandem, progressively through the 7th and 8th centuries. In Chapter 4, I proposed using the average density of charred plant remains in grain-rich samples as a proxy for the magnitude of cereal surplus production and processing. I found by this measure that a considerable increase in surpluses may be dated to the 8th and 9th centuries, whereas the tripartite diversification patterns begin to emerge from the 7th century onwards. We should recall here that there is no absolute, clear-cut distinction between the 'Intermediate' and 'Mid Saxon' phases employed in this book: in some cases, Intermediate and Mid Saxon samples could theoretically be contemporary. Nonetheless, from the data assembled here, it could be argued that the growth of surpluses evident in the 8th and 9th centuries was the result of a process of agricultural development than began earlier, in the 7th century.

As I have argued elsewhere (McKerracher 2018: 121–124), stirrings of agricultural development in the 7th century would have predated the main intensification of craft and trade activity in the early and mid-8th centuries, but would have coincided more nearly with the processes often described as 'kingdom formation': the consolidation of political structures embedded within the landscape, evident in the high-status complexes of the late 6th and early 7th centuries, and later manifested in royal and aristocratic monastic complexes in possession of vast tracts of land. It may therefore be tempting to hold these newcome lords responsible, directly or indirectly, for the agricultural developments of the period – which ultimately, after all, supplied their food rents. On the other hand, correlation does not necessarily mean causation. The Mid Saxon period is not considered to be an era of strong, coercive lordship, and we are inevitably poorly informed as to what motivated the economic activities of the peasants who actually cultivated the crops. It is plausible that the processes of kingdom formation and agricultural development were related, but the nature of that relationship – causal or otherwise – remains to be more deeply explored.

However, such considerations take us beyond the immediate focus of this book, which has been to outline a suite of quantitative and semi-quantitative archaeobotanical methods, and to apply these to a well-defined dataset in order to elucidate a specific period of agricultural change. If some of these mathematically-based approaches seem somewhat sterile, restrictive or simply unimaginative, then those are perhaps valid criticisms. But this formulaic approach is not meant as a substitute for more subtle, qualitative interpretations of archaeobotanical data with regard to past environments and crop husbandry regimes. It is intended to complement such studies, by providing a model for handling datasets too large to interpret without standardised methodologies. It is also meant to support and encourage replication, so that analyses can be performed and updated, and the resultant models rigorously tested, in a more objective way. To this end, I have grafted various methods cultivated by others onto a core, replicable process. Such a process might lend itself well to a degree of computerised automation, a possibility which at the time of writing is being explored as part of the 'Feeding Anglo-Saxon England' project (Hamerow 2017).

In conclusion, this book has argued that, through the combination of quantitative and semi-quantitative methods, data pertaining to Anglo-Saxon crops and weeds can be more than the sum of their parts. By these numbers may we paint an ever more detailed picture of the Anglo-Saxon fields, and of the people who worked that land.

Appendix 1: Key Parameters

Parameter 1: Principal means of preservation [70%]

A sample's principal means of preservation is that which accounts for ≥70% of its whole plant parts.

Parameter 2: Presence analysis quorum [1]

In order to be included in a presence analysis, a sample must contain at least one item (whether whole or fragmentary).

Parameter 3: Dominant crop type [80%]

A crop type is considered dominant in a sample if its standard parts constitute at least 80% of all standard whole plant parts belonging to crop taxa in that sample.

Parameter 4: Quorum for dominance calculation [30]

In order for a dominant crop type to be calculated for a sample, that sample must contain at least 30 standard whole plant parts belonging to crop taxa.

Parameter 5: Quorum for crop processing analysis, by composition ratios [30]

In order for a sample to be included in crop processing analyses based upon the ratios of free-threshing cereal grain, rachis segments, and weed seeds, that sample must contain at least 30 of those items in total.

Parameter 6: Quorum for crop processing discriminant analysis [10]

In order for a sample to be included in crop processing discriminant analyses based upon weed seed types, that sample must contain at least ten weed seeds classified to one of Jones' six types.

Parameter 7: Quorum for calculating average density [30]

Average density was calculated only for those samples with an abundance of at least 30 whole standard plant parts.

Parameter 8: Quorum units for presence analysis [10]

Presence analysis was only conducted for a region or period, if that region or period contained at least ten units of analysis (i.e. assemblages or samples).

Parameter 9: Quorum for relative proportions of grain [30]

A sample must contain at least 30 cereal grains before the relative proportions of different cereal species may be calculated.

Parameter 10: Quorum for correspondence analysis [10]

A sample must contain at least ten weed seeds for it to be included in a weed ecological correspondence analysis. This criterion is applied recursively in tandem with Parameter 11.

Parameter 11: Minimum presence for correspondence analysis [3]

A species must be present in at least three samples for it to be included in a weed ecological correspondence analysis. This criterion is applied recursively in tandem with Parameter 10.

Parameter 12: Significant change for weed presence analysis [10%]

Diachronic change in the presence of a weed species is considered to be significant if it exceeds ten percent, whether in terms of assemblages or samples.

Appendix 2: Key Metadata

Metadata 1: Standardised feature types

ditch/gully
channel
ditch
ditch base
ditch fill
enclosure ditch
gully
linear feature
hearth/oven
hearth
hearth in SFB
oven
other
cut feature
4-post structure
bank
building
clamp
clay
cut
dark earth over ditches
dark earth over quarry pit
depression
dry feature fill
feature
floor
furnace upper fill
furnace rake-out
layer
midden
mound/peat
occupation layer / grid-square
pit/ditch
post-trench
slot
trench

pit/well
lower pit
pit
pit in SFB
refuse pit
shallow feature/pit
well
well basal fill
well/grain storage pit
posthole
fill of ph
fill of ph within SFB
fill of ph/pit
ph
posthole
posthole in SFB
sfb
sunken building
floor of SFB
lower fill of SFB
occupation layer SFB
SFB
SFB backfill
SFB fill
sunken-feature

Metadata 2: Standardised plant parts

culm node
basal culm node
culm base
culm node
floret base
floret base
oat grain with floret base (also counted as **grain**)
glume base
glume
glume base
spikelet base (counted as two **glume bases**)
spikelet fork (counted as two **glume bases**)
rachis
basal internode
basal node
basal rachis
brittle rachis
rachis
rachis basal node
rachis internode
rachis node
tough rachis
tough rachis internode base
grain
caryopsis
grain
oat grain with floret base (also counted as **floret base**)
tail grain
seed
achene
cotyledon (numbers halved, rounded to the nearest integer, and thus counted as **seeds**)
cypsela
kernel
nutlet
seed
other (i.e. non-standard parts)
moss
seed case

vegetative
achenes & seedhead frag.
awn
bud
buds/thorn bases/twigs
capsule
capsule lids
capsule tip
capsule with seeds
catkin
chaff
coleoptile
culm
culm fragment/base
culm internode
embryo
embryo sprout
endocarp
floret
flowering stem
fruit
fruit lids
fruit tissue
glume grot
gristed caryopsis
herbage
hilum
inflorescence
leaf
lemma base
lemma/palea
long awn
milled grain
moss
nutshell
pad
parenchyma
pinnule
plumule
pod

rhizome
root
root/culm
root/rhizome/stem
root/stem
sclerotia/tuber
sclerotium
seedhead
shoot
short awn / long glume
siliqua
siliqua joints
stem
stem/leaf
straw
testa
thorn
tuber
tubers/fruits
various

Metadata 3: Amalgamated plant taxa

Anagallis arvensis L.
Anagallis arvensis L.
Anagallis L.
Aphanes arvensis L. / *australis* Rydb.
Aphanes arvensis L.
Aphanes arvensis L. / *australis* Rydb.
Arrhenatherum elatius (L.) P. Beauv ex. J. & C. Presl
Arrhenatherum (P. Beauv.)
Arrhenatherum elatius (L.) P. Beauv ex. J. & C. Presl
Atriplex patula L. / *prostrata* (Boucher ex. DC.)
Atriplex L.
Atriplex patula L.
Atriplex patula L. / *prostrata* (Boucher ex. DC.)
Avena sativa L.
Avena sativa L.
Avena sativa L. / *strigosa* (Schreb.)
Bromus hordeaceus L. / *secalinus* L.
Bromus hordeaceus L.
Bromus hordeaceus L. / *secalinus* L.
Bromus L.
Bromus secalinus L.
Bupleurum rotundifolium L.
Bupleurum L.
Bupleurum rotundifolium L.
Camelina sativa (L.) Crantz
Camelina (Crantz)
Camelina sativa (L.) Crantz
Cerastium L. / *Stellaria media* (L.) Vill.
Cerastium L. / *Stellaria* L.
Cerastium L. / *Stellaria media* (L.) Vill.
Cereal indet.
Cereals
Secale cereale / *Triticum dicoccum* / *spelta*
Secale / *Hordeum*
Triticum / *Hordeum*
Triticum / *Secale*
Crataegus monogyna (Jacq.)
Crataegus L.
Crataegus monogyna (Jacq.)

Eleocharis palustris (L.) Roem. & Schult. / *uniglumis* (Link) Schult.
Eleocharis (R. Br.)
Eleocharis palustris (L.) Roem. & Schult.
Eleocharis palustris (L.) Roem. & Schult./*uniglumis* (Link) Schult.
Euphrasia L./*Odontites vernus* (Bellardi) Dumort.
Euphrasia L. / *Odontites* (Ludw.)
Euphrasia L. / *Odontites vernus* (Bellardi) Dumort.
Fallopia convolvulus (L.) Á. Löve
Fallopia (Adans.)
Fallopia convolvulus (L.) Á. Löve
Galeopsis tetrahit L.
Galeopsis L.
Galeopsis tetrahit L.
Iris pseudacorus L.
Iris L.
Iris pseudacorus L.
Lepidium campestre (L.) W.T. Aiton
Lepidium campestre (L.) W.T. Aiton
Lepidium L.
Malus sylvestris (L.) Mill./*pumila* (Mill.)
Malus sylvestris (L.) Mill.
Malus sylvestris (L.) Mill./*pumila* (Mill.)
Malva sylvestris L.
Malva L.
Malva sylvestris L.
Malvaceae
Mentha arvensis L. / *aquatica* L.
Mentha arvensis L. / *aquatica* L.
Mentha L.
Montia fontana L.
Montia fontana ssp *chondrosperma* (Fenzl) Walters
Montia fontana ssp *fontana* L.
Myosotis arvensis (L.) Hill
Myosotis arvensis (L.) Hill
Myosotis L.
Odontites vernus (Bellardi) Dumort.
Odontites (Ludw.)
Odontites vernus (Bellardi) Dumort.
Phragmites australis (Cav.) Trin. Ex Steud.
Phragmites (Adans.)
Phragmites australis (Cav.) Trin. Ex Steud.

Plantago lanceolata L.
Plantago lanceolata L.
Plantago media L. / *lanceolata* L.
Poa pratensis L. / *trivialis* L.
Poa pratensis L.
Poa pratensis L. / *trivialis* L.
Prunus domestica L.
Prunus domestica L.
Prunus domestica ssp. *insititia* (L.) Bonnier & Layens
Reseda luteola L.
Reseda L.
Reseda luteola L.
Rhinanthus minor L.
Rhinanthus L.
Rhinanthus minor L.
Schoenoplectus lacustris (L.) Palla
Schoenoplectus (Rchb.) Palla
Schoenoplectus lacustris (L.) Palla
Scleranthus annuus L.
Scleranthus annuus L.
Scleranthus L.
Sinapis alba L. / *arvensis* L.
Sinapis alba L.
Sinapis alba L. / *arvensis* L.
Sinapis arvensis L.
Solanum nigrum L.
Solanum L.
Solanum nigrum L.
Sparganium erectum L.
Sparganium erectum L.
Sparganium L.
Stachys sylvatica L.
Stachys L.
Stachys sylvatica L.
Stellaria graminea L.
Stellaria graminea L.
Stellaria graminea L. / *palustris* (Ehrh. Ex Hoffm.)
Torilis arvensis (Huds.) Link/*japonica* (Houtt.) DC.
Torilis arvensis (Huds.) Link/*japonica* (Houtt.) DC.
Torilis japonica (Houtt.) DC.

Tripleurospermum maritimum (L.) W.D.J. Koch / *inodorum* (L.) Sch. Bip.
Tripleurospermum (Sch. Bip.)
Tripleurospermum inodorum (L.) Sch. Bip.
Tripleurospermum maritimum (L.) W.D.J. Koch / *inodorum* (L.) Sch. Bip.
Triticum L. free-threshing
Triticum aestivo-compactum
Triticum aestivum
Triticum aestivum L. / *compactum* Host.
Triticum aestivum sbsp. *vulgare*
Triticum aestivum / *durum*
Triticum aestivum / *turgidum*
Triticum sp. free-threshing
Triticum turgidum
Vicia faba L.
Vicia faba L.
Vicia faba L. var. *minor*

Metadata 4: Weed seed type classifications

taxon	seed classification
Agrostemma githago L.	bfh
Bromus hordeaceus L./*secalinus* L.	bfh
Convolvulus arvensis L.	bfh
Fabaceae (large)	bfh
Fallopia convolvulus (L.) Á. Löve	bfh
Fumaria officinalis L.	bfh
Galeopsis tetrahit L.	bfh
Galium aparine L.	bfh
Galium palustre L.	bfh
Large legume indet.	bfh
Lathyrus aphaca L.	bfh
Lathyrus L.	bfh
Lolium L.	bfh
Lolium temulentum L.	bfh
Poaceae (large)	bfh
Ranunculus arvensis L./*parviflorus* L.	bfh
Ranunculus subg. *ranunculus* L.	bfh
Torilis arvensis (Huds.) Link/*japonica* (Houtt.) DC.	bfh
Vicia L.	bfh
Vicia sativa L.	bfh
Vicia tetrasperma (L.) Schreb.	bfh
Anthemis arvensis L.	bhh
Anthemis cotula L.	bhh
Avena sterilis L.	bhh
Cladium mariscus (L.) Pohl	bhh
Medicago L.	bhh
Raphanus raphanistrum L.	bhh
Agrostis L./*Poa* L.	sfh
Amaranthaceae	sfh
Aphanes arvensis L./*australis* Rydb.	sfh
Atriplex patula L./*prostrata* (Boucher ex. DC.)	sfh
Bolboschoenus maritimus (L.) Palla	sfh
Brassica L.	sfh
Brassica L./*Sinapis* L.	sfh
Brassica nigra (L.) W.D.J. Koch	sfh
Brassica rapa ssp *campestris* (L.) A.R. Clapham	sfh
Camelina sativa (L.) Crantz	sfh
Carex distans L./*sylvatica* (Huds.)/*laevigata* (Sm.)	sfh

Carex flava L.	sfh
Carex L.	sfh
Chenopodium album L.	sfh
Chenopodium ficifolium (Sm.)	sfh
Chenopodium hybridum L.	sfh
Chenopodium L.	sfh
Chenopodium polyspermum L.	sfh
Chenopodium rubrum L./*glaucum* L.	sfh
Danthonia decumbens (L.) DC.	sfh
Eleocharis palustris (L.) Roem. & Schult./*uniglumis* (Link) Schult.	sfh
Fabaceae (small)	sfh
Hyoscyamus niger L.	sfh
Lolium perenne L.	sfh
Montia fontana L.	sfh
Persicaria lapathifolia (L.) Delarbre	sfh
Persicaria maculosa (Gray)	sfh
Persicaria maculosa (Gray)/*lapathifolia* (L.) Delarbre	sfh
Phleum L.	sfh
Phleum L./*Poa* L.	sfh
Phleum pratense L.	sfh
Plantago major L.	sfh
Poa annua L.	sfh
Poa L.	sfh
Poa pratensis L./*trivialis* L.	sfh
Poaceae (small)	sfh
Polygonum aviculare L.	sfh
Polygonum L.	sfh
Prunella vulgaris L.	sfh
Ranunculus flammula L.	sfh
Rumex acetosa L.	sfh
Rumex acetosella L.	sfh
Rumex conglomeratus (Murray)/*obtusifolius* L./*sanguineus* L.	sfh
Rumex crispus L.	sfh
Rumex L.	sfh
Scirpus L.	sfh
Sherardia arvensis L.	sfh
Sinapis alba L./*arvensis* L.	sfh
Spergula arvensis L.	sfh
Stellaria media (L.) Vill.	sfh
Trifolium arvense L./*campestre* (Shreb.)/*dubium* (Sibth.)/*repens* L.	sfh
Trifolium L.	sfh

Trifolium pratense L.	sfh
Trifolium pratense L./*repens* L.	sfh
Urtica urens L.	sfh
Valerianella dentata (L.) Pollich	sfh
Veronica arvensis L.	sfh
Anisantha sterilis (L.) Nevski	sfl
Euphrasia L./*Odontites vernus* (Bellardi) Dumort.	sfl
Juncus L.	sfl
Odontites vernus (Bellardi) Dumort.	sfl
Sonchus asper (L.) Hill	sfl
Tripleurospermum inodorum (L.) Sch. Bip.	sfl
Arenaria serpyllifolia L.	shh
Glebionis segetum (L.) Fourr.	shh
Malva sylvestris L.	shh
Medicago lupulina L.	shh
Plantago lanceolata L.	shh
Silene flos-cuculi (L.) Clairv.	shh
Silene L.	shh
Silene latifolia (Poir.)	shh
Hordeum murinum L.	shl
Papaver rhoeas L./*dubium* L.	shl

Metadata 5: Flowering habit classifications

Data are given for annual weed taxa only.

taxon	flowering onset / duration
Agrostemma githago L.	early / short
Anagallis arvensis L.	early / short
Anthemis cotula L.	late
Atriplex patula L./*prostrata* (Boucher ex. DC.)	intermediate
Brassica nigra (L.) W.D.J. Koch	early / short
Bromus hordeaceus L./*secalinus* L.	early / short
Chenopodium album L.	late
Fallopia convolvulus (L.) Á. Löve	late
Galium aparine L.	intermediate
Lapsana communis L.	intermediate
Lathyrus nissolia L.	early / short
Lithospermum arvense L.	early / short
Lolium temulentum L.	early / short
Papaver argemone L.	early / short
Persicaria maculosa (Gray)	intermediate
Polygonum aviculare L.	late
Scleranthus annuus L.	early / short
Spergula arvensis L.	early / short
Stellaria media (L.) Vill.	long
Thlaspi arvense L.	early / short
Urtica urens L.	intermediate
Vicia tetrasperma (L.) Schreb.	intermediate

Appendix 3: Gazetteer of Sites

site name	county	eastings	northings	National Character Area	references
Alchester (extramural)	Oxon	457155	220955	Upper Thames Clay Vales	Pelling in Booth *et al.* 2001: 418–422
Ashwell Site (West Fen Road, Ely)	Cambs	552955	280855	The Fens	Ballantyne in Mortimer *et al.* 2005: 100–112
Bancroft	Bucks	482735	240335	Bedfordshire and Cambridgeshire Claylands	Nye and Jones in Williams and Zeepvat 1994: 562–565
Barrow Hills	Oxon	451355	198155	Upper Thames Clay Vales	Moffett in Chambers and McAdam 2007: 290–295
Barton Court Farm	Oxon	451055	197855	Upper Thames Clay Vales	Jones in Miles 1986: microfiche 9:A1-B5
Benson (St Helen's Avenue)	Oxon	461590	191550	Upper Thames Clay Vales	Pine and Ford 2003; Robinson n.d.
Berinsfield (Mount Farm)	Oxon	458225	196615	Upper Thames Clay Vales	Jones in Lambrick 2010: Appendix 19
Bishop's Cleeve	Glos	395855	227560	Cotswolds	Pelling cited in Lovell *et al.* 2007: 111–118
Bletchley (Water Eaton)	Bucks	488055	232625	Bedfordshire and Cambridgeshire Claylands	Martin cited in Hancock 2010: 305–316
Bloodmoor Hill	Suffolk	651855	289955	Suffolk Coast and Heaths	Ballantyne in Lucy *et al.* 2009 : 305–316
Brandon (Staunch Meadow)	Suffolk	577855	286455	Breckland	Murphy and Fryer in Tester *et al.* 2014: 313–330
Brandon Road North (Thetford)	Norfolk	585555	283255	Breckland	Fryer in Atkins and Connor 2010: 102–105
Brettenham (Melford Meadows)	Norfolk	587855	282655	Breckland	Robinson in Mudd 2002: 108–110
Chadwell St Mary (County Primary School)	Essex	564505	178545	Northern Thames Basin	Fryer and Murphy in Lavender 1998: 54–56
Cherry Orton Road (Orton Waterville)	Cambs	515685	296275	Bedfordshire and Cambridgeshire Claylands	Stevens in Wright 2004: 9–11
Chiefs Street (Ely)	Cambs	553565	280425	The Fens	Stevens in Kenney 2002

site name	county	eastings	northings	National Character Area	references
Chiefs Street (Ely)	Cambs	553565	280425	The Fens	Stevens in Kenney 2002
Chieveley	Berks	447955	172755	Thames Basin Heaths	Fryer in Mudd 2007: 76–81
Childerley Gate	Cambs	535955	259855	Bedfordshire and Cambridgeshire Claylands	Giorgi in Abrams and Ingham 2008. CD-ROM Appendix 15
Cogges (Manor Farm)	Oxon	436215	209635	Upper Thames Clay Vales	Robinson in Rowley and Steiner 1996: 133
Collingbourne Ducis (Cadley Road)	Wilts	424450	154000	Salisbury Plain and West Wiltshire Downs	Letts in Pine 2001: 112–113
Consortium Site (West Fen Road, Ely)	Cambs	553155	280955	The Fens	Carruthers in Mudd and Webster 2011: 110–116
Cottenham (Lordship Lane)	Cambs	544955	268155	Bedfordshire and Cambridgeshire Claylands	Stevens in Mortimer 1998
Cresswell Field	Oxon	447055	211455	Upper Thames Clay Vales	Pelling in Hey 2004: 367–368
Criminology Site (Cambridge)	Cambs	544285	258125	Bedfordshire and Cambridgeshire Claylands	Roberts cited in Dodwell et al. 2004: 119–120
Didcot (Milton Park)	Oxon	449705	192355	Upper Thames Clay Vales	Evans in Williams 2008: 23–24
Dumbleton (Bank Farm)	Glos	402725	236585	Severn and Avon Vales	Carruthers in Coleman et al. 2006 : 78–87
Eaton Socon (Alpha Park)	Cambs	516855	258155	Bedfordshire and Cambridgeshire Claylands	Smith in Hood 2007: 64–66
Eye (Hartismere High School)	Suffolk	613800	274040	South Norfolk and High Suffolk Claylands	Caruth 2008; Fryer 2008
Eynesbury	Cambs	518055	258555	Bedfordshire and Cambridgeshire Claylands	Clapham in Ellis 2004: 71–79
Eynsham (Eynsham Abbey)	Oxon	443355	209155	Upper Thames Clay Vales	Pelling in Hardy et al. 2003: 439–448
Flixton (Flixton Park Quarry)	Suffolk	630370	286670	South Norfolk and High Suffolk Claylands	Fryer in Boulter 2008: 220–233
Forbury House (Reading)	Berks	471805	173505	Thames Valley	Edwards 2008; Vaughan-Williams 2005

site name	county	eastings	northings	National Character Area	references
Fordham (Hillside Meadow)	Cambs	563255	270655	East Anglian Chalk	Smith in Patrick and Rátkai 2011: 95–102
Gamlingay	Cambs	524305	251905	Bedfordshire Greensand Ridge	Fryer in Murray 2005: 248–251
Godmanchester	Cambs	525505	270305	Bedfordshire and Cambridgeshire Claylands	Fryer in Gibson 2003: 197–199
Goring (Gatehampton Farm)	Oxon	460655	179755	Chilterns	Letts in Allen et al. 1995: 107–109
Great Linford (Church)	Bucks	485655	241955	Bedfordshire and Cambridgeshire Claylands	Busby in Mynard and Zeepvat 1992: 230–231
Handford Road (Ipswich)	Suffolk	615305	244555	South Suffolk and North Essex Clayland	Fryer in Boulter 2005: 85–96
Harston Mill	Cambs	541855	250755	East Anglian Chalk	Scaife in O'Brien 2016: 187–192
Hintlesham (Silver Birches)	Suffolk	609255	243405	South Suffolk and North Essex Clayland	Fryer in Boulter 2010: 25–27
Hinxton Quarry	Cambs	548755	246655	East Anglian Chalk	Stevens in Mortimer and Evans 1996
Hungerford (Charnham Lane)	Berks	433555	169255	Berkshire and Marlborough Downs	Carruthers in Ford 2002: 54–63
Hutchison Site (Addenbrooke's, Cambridge)	Cambs	546225	255355	East Anglian Chalk	Roberts in Evans et al. 2008: 110–122
Ingleborough (West Walton)	Norfolk	547275	314815	The Fens	Murphy in Crowson et al. 2005: 238–260
Ipswich (various urban excavations)	Suffolk	616255	244155	South Suffolk and North Essex Clayland	Murphy 2004; Suffolk County Council Archaeological Service 2015
Kilverstone	Norfolk	588405	283855	Breckland	Ballantyne in Garrow et al. 2006: 198–199
Lackford Bridge Quarry	Suffolk	579155	271355	Breckland	Tipper 2007; Murphy in West 1985: 100–108
Lake End Road (Dorney)	Bucks	492945	179605	Thames Valley	Pelling in Foreman et al. 2002: 49–55
Latton Quarry	Wilts	408355	195655	Upper Thames Clay Vales	Cramp in Pine 2009: 19–21
Lechlade (Sherborne House)	Glos	421265	199745	Upper Thames Clay Vales	Stevens in Bateman et al. 2003: 76–81

site name	county	eastings	northings	National Character Area	references
Littlemore (Oxford Science Park)	Oxon	453905	202105	Midvale Ridge	Pelling in Moore 2001: 212–213
Lot's Hole (Dorney)	Bucks	492205	179705	Thames Valley	Pelling in Foreman et al. 2002: 49–55
Lower Cambourne	Cambs	531080	259460	Bedfordshire and Cambridgeshire Claylands	Stevens in Wright et al. 2009: 116
Lower Icknield Way	Bucks	489355	212655	Chilterns	Scaife in Masefield 2008: 147–156
Lower Slaughter (Copsehill Road)	Glos	416505	222675	Cotswolds	Jones in Kenyon and Watts 2006: 104–105
Marham (Old Bell)	Norfolk	570845	309795	North West Norfolk	Scaife in Newton 2010: 63–67
Market Lavington	Wilts	401355	154155	Avon Vale	Straker in Williams and Newman 2006: 137–149
Neptune Wood	Oxon	455055	193755	Upper Thames Clay Vales	Pelling in Allen et al. 2010: 237–239
Outwell (Church Terrace)	Norfolk	551455	303755	The Fens	Fryer in Hall 2003: 42–44
Pampisford (Bourn Bridge)	Cambs	551655	249555	East Anglian Chalk	Fryer and Murphy in Pollard 1996
Pennyland	Bucks	486255	241155	Bedfordshire and Cambridgeshire Claylands	Jones in Williams 1993: 171–174
Pitstone	Bucks	493825	215075	Chilterns	Robinson in Phillips 2005: 24–26
RAF Lakenheath	Suffolk	573415	280465	Breckland	Fryer in Caruth 2006: 43–45
Redcastle Furze (Thetford)	Norfolk	586155	283055	Breckland	Murphy in Andrews 1995: 131–135
Rosemary Lane (Cherry Hinton)	Cambs	548555	257655	Bedfordshire and Cambridgeshire Claylands	Roberts in Mortimer 2003
Rycote (Site 30)	Oxon	466095	204955	Upper Thames Clay Vales	Robinson in Taylor and Ford 2004: 31–32
Sedgeford (Chalkpit Field)	Norfolk	571155	336355	North West Norfolk	Fryer in Davies 2008: 191–197
Slough House Farm	Essex	587355	209155	Greater Thames Estuary	Murphy in Wallis and Waughman 1998: 196–204
Spong Hill	Norfolk	598155	319555	Mid Norfolk	Murphy in Rickett 1995: 140–141
Spring Road (Abingdon)	Oxon	448755	197555	Upper Thames Clay Vales	Robinson in Allen and Kamash 2008: 58–59

site name	county	eastings	northings	National Character Area	references
Stansted Airport	Essex	552355	222455	South Suffolk and North Essex Clayland	Murphy in Havis and Brooks 2004: 65–68
Stonea Grange	Cambs	544955	293755	The Fens	van der Veen in Jackson and Potter 1996: 613–629
Sutton Courtenay (Drayton Road)	Oxon	449000	193600	Upper Thames Clay Vales	Robinson cited in Hamerow et al. 2007: 157–160
Taplow Court	Bucks	490755	182355	Thames Valley	Robinson in Allen et al. 2009: 149-151
Terrington St Clement	Norfolk	553865	318005	The Fens	Murphy in Crowson et al. 2005: 238–260
Two Mile Bottom (Thetford)	Norfolk	585255	286855	Breckland	Murphy in Bates and Lyons 2003: 91–93
Walpole St Andrew	Norfolk	548745	316005	The Fens	Murphy in Crowson et al. 2005: 238–260
Walton Lodge (Aylesbury)	Bucks	482385	213245	Upper Thames Clay Vales	Giorgi in Dalwood et al. 1989: 184–185
Walton Orchard (Aylesbury)	Bucks	482385	213285	Upper Thames Clay Vales	Letts in Ford and Howell 2004: 84–85
Walton Road Stores (Aylesbury)	Bucks	482455	213355	Upper Thames Clay Vales	Bonner 1997; Robinson 1997
Walton Street (82-84 Walton Street, Aylesbury)	Bucks	482255	213255	Upper Thames Clay Vales	Livarda in Stone 2011: 122–125
Walton Vicarage (Aylesbury)	Bucks	482255	213255	Upper Thames Clay Vales	Monk in Farley 1976: 171–173
Waterbeach (Denny End)	Cambs	549355	265725	The Fens	Stevens in Mortimer 1996
Wavendon Gate	Bucks	490355	236955	Bedfordshire and Cambridgeshire Claylands	Letts in Williams et al. 1996: 244–256
West Stow	Suffolk	579705	271355	Breckland	Murphy in West 1985: 100–108
Whitehouse Road (Bramford, Ipswich)	Suffolk	613855	247055	South Suffolk and North Essex Clayland	Caruth 1996; Fryer and Murphy 1996
Wicken Bonhunt	Essex	551155	233555	South Suffolk and North Essex Clayland	Wade 1980; Jones n.d.
Wickhams Field	Berks	467505	169705	Thames Valley	Scaife in Crockett 1996: 157–162

site name	county	eastings	northings	National Character Area	references
Willingham (High Street)	Cambs	540405	270375	Bedfordshire and Cambridgeshire Claylands	Fryer in Fletcher 2008: 79–87
Wilton (Wilton Autos)	Wilts	409420	131370	Salisbury Plain and West Wiltshire Downs	Pelling in De'Athe 2012: 139–140
Wittering (Bonemills Farm)	Cambs	504755	301535	Rockingham Forest	Clapham in Wall 2011: 94–95
Witton	Norfolk	633655	332055	North East Norfolk and Flegg	Jones in Lawson 1983: 67–68
Wolverton Turn	Bucks	480255	240665	Bedfordshire and Cambridgeshire Claylands	Robinson and Letts in Preston 2007: 109
Worton	Oxon	446055	211255	Upper Thames Clay Vales	Robinson in Hey 2004: 368
Wymondham (Browick Road)	Norfolk	612455	301555	South Norfolk and High Suffolk Claylands	Ames 2005
Yarnton	Oxon	447555	211355	Upper Thames Clay Vales	Stevens in Hey 2004: 351–364

Appendix 4: Inventory of Samples

Shaded samples are considered not to be independent, for the purposes of this study.

code	site	feature	chronology	context	sample	volume (litres)	abundance	average density
AL1	Alchester	ditch	Early Saxon	5124	70	5	1075	215.0
ALP1	Eaton Socon	sfb	Early Saxon	1256	13	20	31	1.6
ALP2	Eaton Socon	posthole	Early Saxon	1297	15	10	5	-
BBP1	Pampisford	sfb	Intermediate	447	3	8	-	-
BC1	Bancroft	ditch	Early Saxon	94/95	-	-	4	-
BCF1	Barton Court Farm	pit	Early Saxon	1174	-	20	56	2.8
BCF2	Barton Court Farm	sfb	Early Saxon	1190	-	10	15	-
BCL1	Bishop's Cleeve	ditch	Intermediate	1246	14	30	60	2.0
BCL2	Bishop's Cleeve	ditch	Intermediate	2238	21	15	51	3.4
BCL3	Bishop's Cleeve	ditch	Intermediate	1257	12	30	552	18.4
BEL1-12	Marham	sfb	Mid Saxon	2034	multiple	-	220	-
BEL13+20	Marham	hearth/oven	Mid Saxon	2107	45 & 89	-	153	-
BEL14-19	Marham	sfb	Mid Saxon	2108	multiple	-	219	-
BFD1	Dumbleton	ditch	Generic	1540	1068	40	32	0.8
BH1	Barrow Hills	ditch	Early Saxon	601/5/2	-	28.5	-	-
BH2	Barrow Hills	ditch	Early Saxon	601/5/3	-	0.75	-	-
BH3	Barrow Hills	sfb	Early Saxon	1005D/2	-	15	8	-
BH4	Barrow Hills	sfb	Early Saxon	1053/B/2	-	10	-	-
BH5	Barrow Hills	sfb	Early Saxon	1053/D/2	-	10	-	-
BH6	Barrow Hills	sfb	Early Saxon	1061/A/2	-	11	3	-
BH7	Barrow Hills	sfb	Early Saxon	1105/B/4	-	10	2	-
BH8	Barrow Hills	sfb	Early Saxon	2143/B	-	10	1	-
BH9	Barrow Hills	sfb	Early Saxon	1225/B/1	-	12	1	-
BH10	Barrow Hills	sfb	Early Saxon	1225/D/1	-	11	10	-
BH11	Barrow Hills	sfb	Early Saxon	1297/B/4	-	10	-	-
BH12	Barrow Hills	sfb	Early Saxon	1297/D/4	-	10	-	-
BH13	Barrow Hills	sfb	Early Saxon	3606/D/3	-	10	-	-
BH14	Barrow Hills	sfb	Early Saxon	3606/B/2	-	10	-	-

code	site	feature	chronology	context	sample	volume (litres)	abundance	average density
BH15	Barrow Hills	sfb	Early Saxon	3607/D/1	-	10	4	-
BH16	Barrow Hills	sfb	Early Saxon	3608/L/2	-	10	15	-
BH17	Barrow Hills	sfb	Early Saxon	3805/B/3	-	1.5	-	-
BH18	Barrow Hills	sfb	Early Saxon	3805/D/3	-	1.5	-	-
BH19	Barrow Hills	sfb	Early Saxon	3811/D/1	-	2.5	15	-
BH20	Barrow Hills	sfb	Early Saxon	4198/D/2	-	10	152	15.2
BMH1	Bloodmoor Hill	posthole	Generic	4519	617	30	42	1.4
BMH2	Bloodmoor Hill	posthole	Generic	4520	618	40	80	2.0
BMH3	Bloodmoor Hill	posthole	Generic	4529	619	20	26	-
BMH4	Bloodmoor Hill	posthole	Generic	4538	620	4	4	-
BMH5	Bloodmoor Hill	posthole	Generic	4545	623	10	1	-
BMH6	Bloodmoor Hill	posthole	Generic	4553	626	10	3	-
BMH7	Bloodmoor Hill	posthole	Generic	4555	627	10	1	-
BMH8	Bloodmoor Hill	posthole	Generic	4556	628	10	1	-
BMH9	Bloodmoor Hill	posthole	Generic	4534	629	10	3	-
BMH10	Bloodmoor Hill	posthole	Generic	4570	630	4	1	-
BMH11	Bloodmoor Hill	pit	Generic	3469	534	2	-	-
BMH12	Bloodmoor Hill	pit	Generic	3460	551	5	-	-
BMH13	Bloodmoor Hill	pit	Generic	3536	552	5.5	-	-
BMH14	Bloodmoor Hill	pit	Generic	4600	647	10	43	4.3
BMH15	Bloodmoor Hill	pit	Generic	4604	649	20	99	5.0
BON1	Wittering	other	Mid Saxon	1003	-	-	-	-
BON2	Wittering	other	Mid Saxon	1007	-	-	3	-
BON3	Wittering	pit	Mid Saxon	1013	-	-	-	-
BRT1	Brandon Road North	sfb	Early Saxon	2206	3	-	516	-
BRT2	Brandon Road North	well	Early Saxon	1280	46	-	671	-
BRT3	Brandon Road North	pit	Mid Saxon	278	7	-	198	-
BRT4	Brandon Road North	other	Mid Saxon	2315	47	-	440	-
CD1	Collingbourne Ducis	sfb	Mid Saxon	217	337	-	1	-
CD2	Collingbourne Ducis	sfb	Mid Saxon	217	268	-	3	-
CD3	Collingbourne Ducis	sfb	Mid Saxon	217	273	-	1	-

code	site	feature	chronology	context	sample	volume (litres)	abundance	average density
CD4	Collingbourne Ducis	sfb	Mid Saxon	222	272	-	7	-
CD5	Collingbourne Ducis	sfb	Mid Saxon	329	385	-	2	-
CD6	Collingbourne Ducis	sfb	Early Saxon	220	270	-	-	-
CD7	Collingbourne Ducis	sfb	Early Saxon	220	279	-	-	-
CD8	Collingbourne Ducis	sfb	Early Saxon	307	359	-	1	-
CD9	Collingbourne Ducis	sfb	Mid Saxon	304	357	-	1	-
CD10	Collingbourne Ducis	sfb	Mid Saxon	312	363	-	3	-
CD11	Collingbourne Ducis	sfb	Mid Saxon	318	369	-	1	-
CD12	Collingbourne Ducis	sfb	Mid Saxon	321	273	-	1	-
CD13	Collingbourne Ducis	sfb	Mid Saxon	333	388	-	-	-
CD14	Collingbourne Ducis	sfb	Mid Saxon	333	399	-	-	-
CD15	Collingbourne Ducis	sfb	Mid Saxon	341	398	-	-	-
CD16	Collingbourne Ducis	sfb	Mid Saxon	340	470	-	3	-
CD17	Collingbourne Ducis	sfb	Mid Saxon	349	467	-	3	-
CD18	Collingbourne Ducis	other	Mid Saxon	511	554	-	-	-
CD19	Collingbourne Ducis	other	Mid Saxon	511	556	-	-	-
CD20	Collingbourne Ducis	other	Mid Saxon	200	250	-	3	-
CD21	Collingbourne Ducis	other	Mid Saxon	202	252	-	2	-
CD22	Collingbourne Ducis	other	Mid Saxon	203	254	-	2	-
CD23	Collingbourne Ducis	other	Mid Saxon	206	258	-	3	-
CD24	Collingbourne Ducis	other	Mid Saxon	336	392	-	3	-
CF1	Cresswell Field	sfb	Mid Saxon	7311	7024	37	23	-
CF2	Cresswell Field	sfb	Mid Saxon	7312	7025	37	25	-
CF3	Cresswell Field	sfb	Mid Saxon	7841	7030/1	47	25	-
CF4	Cresswell Field	sfb	Mid Saxon	8556	7102	37	11	-
CHG1	Childerley Gate	other	Early Saxon	group 7.4	5023	10	12	-
CHG2	Childerley Gate	other	Early Saxon	group 14.2	5037	10	34	3.4
CHG3	Childerley Gate	other	Early Saxon	group 14.2	5034	10	2	-
CHM1-3	Chadwell St Mary	sfb	Intermediate	17	multiple	45	433	9.6
CHM4-5	Chadwell St Mary	sfb	Intermediate	12	6 & 7	30	50	1.7

code	site	feature	chronology	context	sample	volume (litres)	abundance	average density
CHM6-7	Chadwell St Mary	sfb	Intermediate	13	8 & 9	30	11	-
CHM8-9	Chadwell St Mary	sfb	Intermediate	19	10 & 11	30	9	-
CHM10	Chadwell St Mary	sfb	Intermediate	21	12	15	11	-
CHM11	Chadwell St Mary	sfb	Intermediate	34	13	15	30	2.0
CHN1	Hungerford	sfb	Generic	SFB 5001	5081	4	3	-
CHO1	Cherry Orton Road	sfb	Intermediate	95	-	30	57	1.9
CHV1	Chieveley	pit	Generic	29218	20	40	-	-
CHV2	Chieveley	pit	Generic	1003	2	20	-	-
CHV3	Chieveley	pit	Generic	1010	3	20	-	-
CHV4	Chieveley	pit	Generic	1018	4	20	-	-
COG1	Cogges	posthole	Early Saxon	141	-	-	-	-
COG2	Cogges	sfb	Early Saxon	146	-	-	-	-
COG3	Cogges	posthole	Early Saxon	148	-	-	-	-
COG4	Cogges	posthole	Early Saxon	152	-	-	-	-
CPS1	Sedgeford	ditch	Mid Saxon	202	-	10	-	-
CPS2	Sedgeford	ditch	Mid Saxon	210	-	10	-	-
CPS3	Sedgeford	ditch	Mid Saxon	230	-	10	-	-
CPS4	Sedgeford	ditch	Mid Saxon	412	-	10	-	-
CPS5	Sedgeford	pit	Mid Saxon	509	-	10	-	-
CPS6	Sedgeford	ditch	Mid Saxon	308	-	20	-	-
CPS7	Sedgeford	ditch	Mid Saxon	310	-	20	-	-
CPS8	Sedgeford	hearth/oven	Mid Saxon	311	-	20	-	-
CPS9	Sedgeford	hearth/oven	Mid Saxon	312	-	20	-	-
CPS10	Sedgeford	hearth/oven	Mid Saxon	315	-	5	-	-
CR1	Lower Slaughter	ditch	Mid Saxon	1081	1	30	28	-
CR2	Lower Slaughter	ditch	Mid Saxon	319	6	30	12	-
CR3	Lower Slaughter	ditch	Mid Saxon	323	8	30	17	-
CR4	Lower Slaughter	ditch	Mid Saxon	130	4	30	27	-
CR5	Lower Slaughter	ditch	Mid Saxon	079	7	30	13	-
CR6	Lower Slaughter	ditch	Mid Saxon	343	9	30	20	-

code	site	feature	chronology	context	sample	volume (litres)	abundance	average density
CR7	Lower Slaughter	ditch	Mid Saxon	272	5	30	51	1.7
CR8	Lower Slaughter	ditch	Mid Saxon	049	3	30	29	-
CRM1	Criminology Site	pit	Early Saxon	10	1	7	13	-
CRM2	Criminology Site	posthole	Early Saxon	66	3	1	9	-
CRM3	Criminology Site	pit	Early Saxon	60	4	2	111	55.5
CRM4	Criminology Site	sfb	Early Saxon	123	5	4	75	18.8
CSE1	Chiefs Street	well	Mid Saxon	102	4	-	34	-
CSE2	Chiefs Street	well	Mid Saxon	97	5	-	54	-
DEW1	Waterbeach	pit	Early Saxon	12	13	15	11	-
DEW2	Waterbeach	pit	Early Saxon	8	1	15	14	-
DEW3	Waterbeach	pit	Early Saxon	18	8	15	5	-
DEW4	Waterbeach	pit	Early Saxon	23	9	15	17	-
DEW5	Waterbeach	pit	Early Saxon	14	6	15	6	-
EN1	Eynsham	pit	Intermediate	394	86	10	4	-
EN2	Eynsham	posthole	Intermediate	1984	176	0.5	24	-
EN3	Eynsham	posthole	Intermediate	3011	177	1	39	39.0
ENB1	Eynesbury	pit	Early Saxon	4605	1560	10	149	14.9
ENB2	Eynesbury	pit	Early Saxon	4606	1562	10	122	12.2
ENB3	Eynesbury	sfb	Early Saxon	7201	9501	30	362	12.1
ENB4	Eynesbury	sfb	Early Saxon	8617	9505	30	39	1.3
ENB5	Eynesbury	posthole	Early Saxon	8612	9506	10	26	-
ENB6	Eynesbury	pit	Early Saxon	7406	9514	1	5	-
FLX1	Flixton	sfb	Early Saxon	276	548	10	-	-
FLX2	Flixton	sfb	Early Saxon	591	592	8	-	-
FLX3	Flixton	sfb	Early Saxon	278	555	16	-	-
FLX4	Flixton	sfb	Early Saxon	605	685	16	-	-
FLX5	Flixton	sfb	Early Saxon	1120	1125	16	-	-
FLX6-7	Flixton	sfb	Early Saxon	1501	1506 & 1520	16	-	-
FLX8-9	Flixton	pit	Early Saxon	604	610 & 746	16	-	-
FLX10	Flixton	pit	Early Saxon	785	786	16	-	-

code	site	feature	chronology	context	sample	volume (litres)	abundance	average density
FLX11	Flixton	pit	Early Saxon	846	847	8	-	-
FLX12	Flixton	pit	Early Saxon	857	858	8	-	-
FLX13-14	Flixton	pit	Early Saxon	1121	1144 & 1153	16	-	-
FLX15	Flixton	pit	Early Saxon	1343	1344	8	-	-
FLX16	Flixton	pit	Early Saxon	1461	1462	8	-	-
FLX17	Flixton	posthole	Early Saxon	226	227	8	-	-
FLX18	Flixton	posthole	Early Saxon	304	305	8	-	-
FLX19	Flixton	posthole	Early Saxon	344	345	8	-	-
FLX20	Flixton	posthole	Early Saxon	631	632	8	-	-
FLX21	Flixton	posthole	Early Saxon	702	703	8	-	-
FLX22	Flixton	posthole	Early Saxon	653	356	8	-	-
FLX23	Flixton	posthole	Early Saxon	374	375	8	-	-
FLX24	Flixton	posthole	Early Saxon	402	403	8	-	-
FLX25	Flixton	posthole	Early Saxon	908	910	8	-	-
FLX26	Flixton	posthole	Early Saxon	1241	1242	8	-	-
FLX27	Flixton	posthole	Early Saxon	1243	1244	8	-	-
FLX28	Flixton	posthole	Early Saxon	1255	1255	8	-	-
FLX29	Flixton	posthole	Early Saxon	1351	1352	8	-	-
FLX30	Flixton	posthole	Early Saxon	1361	1362	8	-	-
FLX31	Flixton	ditch	Early Saxon	314	315	16	-	-
FLX32	Flixton	ditch	Early Saxon	314	687	16	-	-
FLX33	Flixton	ditch	Early Saxon	318	319	16	-	-
FOR1	Forbury House	pit	Mid Saxon	53	1	20	816	40.8
GAM1	Gamlingay	sfb	Intermediate	2077	17	30	49	1.6
GAM2	Gamlingay	pit	Intermediate	2402	56	30	494	16.5
GAM3	Gamlingay	sfb	Intermediate	2525	72	30	18	-
GAM4	Gamlingay	sfb	Intermediate	2562	76	30	114	3.8
GAM5	Gamlingay	ditch	Mid Saxon	2648	78	50	561	11.2
GF1	Goring	other	Early Saxon	421	-	10	8	-
GL1	Great Linford	other	Generic	55	-	2	-	-

code	site	feature	chronology	context	sample	volume (litres)	abundance	average density
GMC1	Godmanchester	other	Mid Saxon	2477	21	-	-	-
GMC2	Godmanchester	other	Mid Saxon	2610	23	-	-	-
GMC3	Godmanchester	other	Mid Saxon	2741	24	-	-	-
GMC4	Godmanchester	sfb	Mid Saxon	2672	26	-	-	-
GMC5	Godmanchester	sfb	Mid Saxon	2703	28	-	-	-
GMC6	Godmanchester	sfb	Mid Saxon	2538	30	-	-	-
GMC7	Godmanchester	sfb	Mid Saxon	2540	33	-	-	-
GMC8	Godmanchester	sfb	Mid Saxon	2705	35	-	-	-
HAM1	Harston Mill	ditch	Generic	L2104	1	20	28	-
HAM2	Harston Mill	pit	Mid Saxon	L2199	3	20	569	28.5
HAM3	Harston Mill	pit	Generic	L2388	5	20	77	3.9
HAM4	Harston Mill	pit	Generic	L2527	6	20	118	5.9
HAM5	Harston Mill	pit	Generic	L2531	4	20	203	10.2
HAM6-7	Harston Mill	pit	Generic	L2556	7A & 7B	40	334	8.4
HAM8	Harston Mill	pit	Mid Saxon	L2665	10	20	79	4.0
HAM9	Harston Mill	pit	Mid Saxon	L3325	18	20	60	3.0
HAM10	Harston Mill	pit	Mid Saxon	L3517	13	20	104	5.2
HAM11	Harston Mill	pit	Mid Saxon	L3565	14	20	140	7.0
HAM12	Harston Mill	pit	Generic	L3589	15	20	1101	55.1
HAM13	Harston Mill	ditch	Mid Saxon	L3754	82	20	16	-
HAM14	Harston Mill	ditch	Generic	L3758	83	20	3	-
HAM15	Harston Mill	pit	Mid Saxon	L4088	35	20	295	14.8
HAM16-17	Harston Mill	sfb	Early Saxon	L4184	41 & 51	40	21	-
HAM18-20	Harston Mill	sfb	Early Saxon	L4530	multiple	60	135	2.3
HAM21	Harston Mill	pit	Generic	L4653	27	20	54	2.7
HAM22	Harston Mill	pit	Generic	L4677	29	20	36	1.8
HAM23	Harston Mill	pit	Mid Saxon	L4663	30	20	15	-
HAM24	Harston Mill	pit	Generic	L4727	34	20	18	-
HAM25	Harston Mill	posthole	Early Saxon	L4431	69	20	8	-
HAM26	Harston Mill	pit	Mid Saxon	L4697	31	20	841	42.1

code	site	feature	chronology	context	sample	volume (litres)	abundance	average density
HAM27	Harston Mill	pit	Generic	L4819	42	20	1021	51.1
HAM28	Harston Mill	pit	Generic	L4895	46	20	311	15.6
HAM29-31	Harston Mill	sfb	Early Saxon	L4933	multiple	60	45	0.8
HAM32	Harston Mill	pit	Mid Saxon	L5436	93	20	383	19.2
HAM33	Harston Mill	pit	Generic	L5460	94	20	1198	59.9
HAM34	Harston Mill	pit	Generic	L5743	97	20	7	-
HAM35	Harston Mill	sfb	Early Saxon	L6086	138	20	6	-
HAM36	Harston Mill	pit	Generic	L5844	102	20	230	11.5
HAM37-40	Harston Mill	sfb	Early Saxon	L5871	multiple	80	39	0.5
HDI1-12	Handford Road	sfb	Early Saxon	SFB 0699	multiple	384	-	-
HDI13-22	Handford Road	sfb	Early Saxon	SFB 3001	multiple	320	-	-
HDI23	Handford Road	posthole	Early Saxon	SFB 3001	3207	16	-	-
HDI24	Handford Road	posthole	Early Saxon	SFB 3001	3259	16	-	-
HDI25-30	Handford Road	sfb	Early Saxon	SFB 3002	multiple	192	-	-
HDI31	Handford Road	posthole	Early Saxon	SFB 3002	3102	32	-	-
HDI32	Handford Road	posthole	Early Saxon	SFB 3002	3211	16	-	-
HDI33	Handford Road	posthole	Early Saxon	SFB 3002	3212	16	-	-
HDI34	Handford Road	sfb	Early Saxon	285	286	32	-	-
HDI35-36	Handford Road	sfb	Early Saxon	985	986 & 2001	64	-	-
HDI37-38	Handford Road	sfb	Early Saxon	2554	2555 & 2556	64	-	-
HDI39-42	Handford Road	pit	Early Saxon	3220	multiple	112	-	-
HDI43	Handford Road	pit	Early Saxon	3484	3485	32	-	-
HDI44	Handford Road	hearth/oven	Early Saxon	3405	3760	32	-	-
HDI45	Handford Road	pit	Early Saxon	3372	3948	32	-	-
HDI46	Handford Road	hearth/oven	Early Saxon	4030	3980	32	-	-
HHS1	Eye	sfb	Early Saxon	15	2	10	-	-
HHS2	Eye	sfb	Early Saxon	33	5	10	-	-
HHS3	Eye	sfb	Early Saxon	85	12	10	-	-
HHS4	Eye	sfb	Early Saxon	86	13	10	-	-
HHS5	Eye	sfb	Early Saxon	89	17	10	-	-

code	site	feature	chronology	context	sample	volume (litres)	abundance	average density
HHS6	Eye	sfb	Early Saxon	95	18	10	-	-
HHS7	Eye	sfb	Early Saxon	96	19	10	-	-
HHS8	Eye	sfb	Early Saxon	99	26	10	-	-
HHS9	Eye	posthole	Early Saxon	125	38	10	-	-
HHS10	Eye	sfb	Early Saxon	2015	14	10	-	-
HHS11	Eye	sfb	Early Saxon	2017	15	10	-	-
HHS12	Eye	sfb	Early Saxon	2018	16	10	-	-
HHS13	Eye	sfb	Early Saxon	2025	20	10	-	-
HHS14	Eye	sfb	Early Saxon	2019	23	10	-	-
HHS15	Eye	posthole	Early Saxon	2044	29	10	-	-
HHS16	Eye	posthole	Early Saxon	2045	30	10	-	-
HHS17	Eye	posthole	Early Saxon	2051	31	10	-	-
HHS18	Eye	posthole	Early Saxon	2047	34	10	-	-
HHS19	Eye	posthole	Early Saxon	2049	35	10	-	-
HHS20	Eye	posthole	Early Saxon	2050	36	10	-	-
HHS21	Eye	sfb	Early Saxon	202	43	10	-	-
HHS22	Eye	sfb	Early Saxon	203	44	10	-	-
HHS23	Eye	sfb	Early Saxon	206	45	10	-	-
HHS24	Eye	sfb	Early Saxon	258	57	10	-	-
HHS25	Eye	sfb	Early Saxon	240	58	10	-	-
HHS26	Eye	sfb	Early Saxon	266	59	10	-	-
HHS27	Eye	sfb	Early Saxon	274	60	10	-	-
HHS28	Eye	ditch	Early Saxon	496	127	10	-	-
HHS29	Eye	ditch	Early Saxon	283	283	10	-	-
HHS30	Eye	other	Early Saxon	2506	351	10	-	-
HHS31	Eye	other	Early Saxon	2555	352	10	-	-
HHS32	Eye	sfb	Early Saxon	565	137	10	-	-
HHS33	Eye	sfb	Early Saxon	566	138	10	-	-
HHS34	Eye	sfb	Early Saxon	330	139	10	-	-
HHS35	Eye	sfb	Early Saxon	329	140	10	-	-

code	site	feature	chronology	context	sample	volume (litres)	abundance	average density
HHS36	Eye	sfb	Early Saxon	328	141	10	-	-
HHS37	Eye	sfb	Early Saxon	831	142	10	-	-
HHS38	Eye	sfb	Early Saxon	331	143	10	-	-
HHS39	Eye	sfb	Early Saxon	567	144	10	-	-
HHS40	Eye	sfb	Early Saxon	869	238	10	-	-
HHS41	Eye	sfb	Early Saxon	890	239	10	-	-
HHS42	Eye	sfb	Early Saxon	871	240	10	-	-
HHS43	Eye	sfb	Early Saxon	784	241	10	-	-
HHS44	Eye	sfb	Early Saxon	785	242	10	-	-
HHS45	Eye	sfb	Early Saxon	916	245	10	-	-
HHS46	Eye	sfb	Early Saxon	917	246	10	-	-
HHS47	Eye	sfb	Early Saxon	914	247	10	-	-
HHS48	Eye	sfb	Early Saxon	915	248	10	-	-
HHS49	Eye	sfb	Early Saxon	714	412	10	-	-
HHS50	Eye	sfb	Early Saxon	715	413	10	-	-
HHS51	Eye	sfb	Early Saxon	716	414	10	-	-
HHS52	Eye	sfb	Early Saxon	717	415	10	-	-
HHS53	Eye	sfb	Early Saxon	849	418	10	-	-
HHS54	Eye	sfb	Early Saxon	846	419	10	-	-
HHS55	Eye	sfb	Early Saxon	3113	569	10	-	-
HHS56	Eye	other	Early Saxon	2607	561	10	-	-
HHS57	Eye	other	Early Saxon	2625	564	10	-	-
HHS58	Eye	other	Early Saxon	3095	566	10	-	-
HHS59	Eye	other	Early Saxon	2871	568	10	-	-
HIL1	Fordham	pit	Mid Saxon	1177	20	20	147	7.4
HIL2	Fordham	sfb	Mid Saxon	1258	21	20	174	8.7
HSW1	Willingham	posthole	Mid Saxon	306	50	10	-	-
HSW2	Willingham	posthole	Mid Saxon	395	52	8	-	-
HSW3	Willingham	posthole	Mid Saxon	356	43	4	-	-
HSW4	Willingham	ditch	Mid Saxon	126	21	20	-	-

code	site	feature	chronology	context	sample	volume (litres)	abundance	average density
HUT1	Hutchison Site	ditch	Mid Saxon	401	4	12	29	-
HUT2	Hutchison Site	ditch	Mid Saxon	451	16	8	24	-
HUT3	Hutchison Site	pit	Mid Saxon	662	6	8	1232	154.0
HUT4	Hutchison Site	posthole	Mid Saxon	3560	170	1	1	-
HUT5	Hutchison Site	well	Mid Saxon	1100/1	35	20	1318	65.9
HUT6	Hutchison Site	well	Mid Saxon	1102/3	36	20	1758	87.9
HUT7	Hutchison Site	well	Mid Saxon	1258	41	4	335	83.8
HUT8	Hutchison Site	well	Mid Saxon	1259	42 & 43	6	3	-
HUT9	Hutchison Site	well	Mid Saxon	1244	46	5	253	50.6
HUT10	Hutchison Site	well	Mid Saxon	1245	47	2	-	-
HUT11	Hutchison Site	well	Mid Saxon	3518	159	3	48	16.0
HXQ1	Hinxton Quarry	sfb	Early Saxon	511	4	15	9	-
IPS1	Ipswich	other	Mid Saxon	0007	-	49	66	1.3
IPS2	Ipswich	pit	Mid Saxon	0052	-	126	419	3.3
IPS3	Ipswich	other	Mid Saxon	0109	-	98	585	6.0
IPS4	Ipswich	other	Mid Saxon	0182	-	24.5	82	3.3
IPS5	Ipswich	pit	Mid Saxon	0190	-	135	251	1.9
IPS6	Ipswich	other	Mid Saxon	0203	-	61	157	2.6
IPS7	Ipswich	pit	Mid Saxon	0192	-	-	2812	-
IPS8	Ipswich	pit	Mid Saxon	0648	-	-	35	-
IPS9	Ipswich	pit	Mid Saxon	1668	-	-	464	-
IPS11	Ipswich	pit	Mid Saxon	0046	-	-	427	-
IPS12	Ipswich	pit	Mid Saxon	0596	-	-	162	-
IPS13	Ipswich	other	Mid Saxon	0113	-	60	20	-
IPS14	Ipswich	other	Mid Saxon	0229	-	65	31	0.5
IPS15	Ipswich	other	Mid Saxon	0369	-	40	48	1.2
IPS16	Ipswich	other	Mid Saxon	0374	-	55	18	-
IPS18	Ipswich	other	Mid Saxon	0537	-	75	57	0.8
IPS19	Ipswich	pit	Mid Saxon	0017	-	-	347	-
IPS20	Ipswich	other	Mid Saxon	0019	-	116	139	1.2

code	site	feature	chronology	context	sample	volume (litres)	abundance	average density
IPS21	Ipswich	pit	Mid Saxon	0028	-	80	420	5.3
IPS22	Ipswich	other	Mid Saxon	0035	-	196	150	0.8
IPS23	Ipswich	pit	Mid Saxon	0038	-	98	122	1.2
IPS25	Ipswich	other	Mid Saxon	0238	-	49	93	1.9
IPS26	Ipswich	other	Mid Saxon	0288	-	135	133	1.0
IPS27	Ipswich	other	Mid Saxon	0293	-	126	55	0.4
IPS28	Ipswich	other	Mid Saxon	0320	-	73	113	1.5
IPS29	Ipswich	other	Mid Saxon	0064	-	49	75	1.5
IPS30	Ipswich	ditch	Intermediate	0134	-	61	123	2.0
IPS31	Ipswich	other	Mid Saxon	0141	-	37	55	1.5
IPS32	Ipswich	other	Mid Saxon	0018	-	122	98	0.8
IPS33	Ipswich	other	Mid Saxon	0026	-	98	46	0.5
IPS34	Ipswich	pit	Mid Saxon	0051	-	147	370	2.5
IPS35	Ipswich	other	Mid Saxon	0090	-	622	23	-
IPS36	Ipswich	other	Mid Saxon	0237	-	435	5	-
IPS37	Ipswich	other	Mid Saxon	0281	-	475	38	0.1
IPS38	Ipswich	other	Mid Saxon	0005	-	122	113	0.9
IPS39	Ipswich	other	Mid Saxon	0020	-	147	116	0.8
KLV1	Kilverstone	sfb	Early Saxon	1220	25	13	8	-
KLV2	Kilverstone	sfb	Early Saxon	1457	37	11	3	-
KLV3	Kilverstone	sfb	Early Saxon	3277	79	14	4	-
KLV4-5	Kilverstone	sfb	Early Saxon	3092 & 3094	70 & 71	30	9	-
KLV6-7	Kilverstone	sfb	Early Saxon	3095 & 3096	67 & 72	20	14	-
KLV8-9	Kilverstone	sfb	Early Saxon	705	35 & 36	26	10	-
KLV10-11	Kilverstone	sfb	Early Saxon	3391 & 3394	80 & 83	28	17	-
KLV12	Kilverstone	sfb	Early Saxon	3406	85	13	3	-
KLV13-14	Kilverstone	sfb	Early Saxon	2078 & 2079	38 & 39	23	28	-
LAK1	RAF Lakenheath	ditch	Mid Saxon	92	-	20	-	-
LAK2	RAF Lakenheath	pit	Mid Saxon	171	-	20	-	-
LAK3	RAF Lakenheath	other	Mid Saxon	794	-	20	-	-

code	site	feature	chronology	context	sample	volume (litres)	abundance	average density
LAK4	RAF Lakenheath	ditch	Mid Saxon	809	-	10	-	-
LBQ1-3	Lackford Bridge Quarry	pit	Mid Saxon	026	3 depths	11.55	561	48.6
LBQ4	Lackford Bridge Quarry	sfb	Mid Saxon	101	-	3.85	19	-
LBQ5	Lackford Bridge Quarry	pit	Mid Saxon	143	-	3.85	25	-
LCM1	Lower Cambourne	pit	Generic	5247	560	10	7	-
LCM2	Lower Cambourne	ditch	Generic	5722	578	10	20	-
LE1	Lake End Road	pit	Early Saxon	3476	3016	40	189	4.7
LE2	Lake End Road	pit	Mid Saxon	30589	3030	20	41	2.1
LE3	Lake End Road	pit	Mid Saxon	30231	3008	20	651	32.6
LE4	Lake End Road	pit	Mid Saxon	30230	3007	20	86	4.3
LE5	Lake End Road	pit	Mid Saxon	30266	3005	20	235	11.8
LE6	Lake End Road	pit	Mid Saxon	30659	3046	55	1246	22.7
LE7	Lake End Road	pit	Mid Saxon	30695	3047	60	1586	26.4
LE8	Lake End Road	pit	Mid Saxon	30623	3040	15	51	3.4
LH1	Lot's Hole	pit	Mid Saxon	50310	77	22	918	41.7
LH2	Lot's Hole	pit	Mid Saxon	50284	76	33	201	6.1
LH3	Lot's Hole	pit	Mid Saxon	50287	12	44	120	2.7
LH4	Lot's Hole	pit	Mid Saxon	50815	102	34	104	3.1
LI1	Lower Icknield Way	pit	Intermediate	168	26	40	3	-
LI2	Lower Icknield Way	pit	Intermediate	179	29	40	2	-
LI3	Lower Icknield Way	other	Intermediate	430	107	40	15	-
LI4	Lower Icknield Way	pit	Intermediate	896	71	40	5	-
LI5	Lower Icknield Way	pit	Intermediate	906	78	40	5	-
LLC1	Cottenham	other	Intermediate	118	1	10	12	-
LLC2	Cottenham	ditch	Intermediate	800	69	12	31	2.6
LLC3	Cottenham	other	Intermediate	656	57	10	118	11.8
LLC4	Cottenham	ditch	Intermediate	771	66	9	55	6.1
LLC5	Cottenham	ditch	Intermediate	?	105	5	18	-
LLC6	Cottenham	other	Intermediate	592	53	-	32	-
LLC7	Cottenham	other	Intermediate	661	58	10	33	3.3

code	site	feature	chronology	context	sample	volume (litres)	abundance	average density
LLC8	Cottenham	pit	Intermediate	958	84	6	4	-
LLC9	Cottenham	ditch	Intermediate	1189	110	8	12	-
LLC10	Cottenham	ditch	Intermediate	1135	98	8	152	19.0
LLC11	Cottenham	other	Intermediate	1187	109	15	114	7.6
LLC12	Cottenham	well	Intermediate	1057	90	8	5	-
LLC13	Cottenham	ditch	Intermediate	1066	93	8	16	-
LLC14	Cottenham	ditch	Intermediate	1092	97	8	9	-
LLC15	Cottenham	pit	Intermediate	1125	96	6	1	-
LLC16	Cottenham	pit	Intermediate	1186	108	9	15	-
LLC17	Cottenham	posthole	Intermediate	1242	115	10	42	4.2
LLC18	Cottenham	other	Intermediate	1342	127	11	16	-
LLC19	Cottenham	posthole	Intermediate	1348	128	8	4	-
LLC20	Cottenham	posthole	Intermediate	1355	126	8	29	-
LLC21	Cottenham	other	Mid Saxon	116	4	10	49	4.9
LLC22	Cottenham	other	Mid Saxon	183	6	10	131	13.1
LLC23	Cottenham	other	Mid Saxon	367	34	10	12	-
LLC24	Cottenham	other	Mid Saxon	419	32	10	-	-
LLC25	Cottenham	other	Mid Saxon	?	59	10	-	-
LLC26	Cottenham	other	Mid Saxon	308	24	10	1	-
LLC27	Cottenham	other	Mid Saxon	?	35	10	60	6.0
LM1	Littlemore	other	Early Saxon	107	115	10	1	-
LM2	Littlemore	sfb	Early Saxon	125	131	12	2	-
LM3	Littlemore	sfb	Early Saxon	155	157	10	-	-
LM4	Littlemore	sfb	Early Saxon	155	158	10	-	-
LM5	Littlemore	sfb	Early Saxon	138	123	7	1	-
LM6	Littlemore	sfb	Early Saxon	182	170	10	-	-
LM7	Littlemore	sfb	Early Saxon	182	171	10	-	-
LM8	Littlemore	sfb	Early Saxon	187	185	10	1	-
LQ1	Latton Quarry	pit	Early Saxon	431/562	41	20	2	-
LQ2	Latton Quarry	posthole	Early Saxon	2327/2599	521	20	3	-

code	site	feature	chronology	context	sample	volume (litres)	abundance	average density
LW1	Lake End Road	pit	Mid Saxon	41057	80	30	348	11.6
LW2-3	Lake End Road	pit	Mid Saxon	40478	28	30	1554	51.8
LW4	Lake End Road	pit	Mid Saxon	40624	117	40	2825	70.6
LW5	Lake End Road	pit	Mid Saxon	41125	87	40	674	16.9
LW6	Lake End Road	pit	Mid Saxon	41425	97	40	446	11.2
LW7	Lake End Road	pit	Mid Saxon	40787	53	40	1471	36.8
MFB3	Berinsfield	pit	Generic	39	1	40	-	-
MFB4	Berinsfield	pit	Generic	39	2	40	-	-
MFB5	Berinsfield	pit	Generic	39	3	40	-	-
MFB6	Berinsfield	pit	Generic	39	4	40	-	-
MFB7	Berinsfield	pit	Generic	39	6	40	-	-
MFB8	Berinsfield	pit	Generic	41	-	40	2	-
MFB9	Berinsfield	pit	Generic	42	-	40	7	-
MFB10	Berinsfield	well	Generic	82	3	40	53	1.3
MFB11	Berinsfield	well	Generic	82	6	40	22	-
MFB12	Berinsfield	pit	Generic	283	-	21.2	29	-
MFB13	Berinsfield	pit	Generic	324	-	40	24	-
MFB14	Berinsfield	pit	Generic	664	-	40	98	2.5
ML1	Market Lavington	ditch	Intermediate	1029	-	18	5	-
ML2	Market Lavington	ditch	Intermediate	13711	-	10	8	-
ML3	Market Lavington	ditch	Intermediate	13721	-	20	15	-
ML4	Market Lavington	ditch	Intermediate	13726	-	20	124	6.2
ML5	Market Lavington	pit	Intermediate	13730	-	15	18	-
ML6	Market Lavington	pit	Intermediate	13744	-	10	79	7.9
ML7	Market Lavington	sfb	Intermediate	13750	-	20	12	-
ML8	Market Lavington	sfb	Intermediate	13752	-	18	25	-
ML9	Market Lavington	sfb	Intermediate	13791	-	20	40	2.0
ML10	Market Lavington	ditch	Intermediate	3024	-	15	10	-
ML11	Market Lavington	ditch	Intermediate	3026	-	18	5	-
ML12	Market Lavington	ditch	Intermediate	3045	-	15	8	-

code	site	feature	chronology	context	sample	volume (litres)	abundance	average density
ML13	Market Lavington	ditch	Intermediate	15507	-	20	-	-
ML14	Market Lavington	ditch	Intermediate	15507	-	16	-	-
ML15	Market Lavington	ditch	Intermediate	15507	-	15	-	-
ML16	Market Lavington	ditch	Intermediate	15507	-	18	-	-
ML17	Market Lavington	ditch	Intermediate	15507	-	10	-	-
ML18	Market Lavington	other	Intermediate	14010	-	20	11	-
ML19	Market Lavington	other	Intermediate	14110	-	20	6	-
ML20	Market Lavington	other	Intermediate	14210	-	20	5	-
ML21	Market Lavington	other	Intermediate	14310	-	20	13	-
ML22	Market Lavington	other	Intermediate	14410	-	20	9	-
ML23	Market Lavington	other	Intermediate	14510	-	20	3	-
ML24	Market Lavington	other	Intermediate	14610	-	20	4	-
ML25	Market Lavington	other	Intermediate	14710	-	20	5	-
MMB1	Brettenham	hearth/oven	Early Saxon	2165	5	12	355	29.6
MMB2	Brettenham	sfb	Early Saxon	2280	16	12	11	-
MMB3	Brettenham	pit	Early Saxon	2430	17	12	13	-
MMB4	Brettenham	sfb	Early Saxon	2536	21	12	33	2.8
MMB5	Brettenham	pit	Early Saxon	2850	24	12	3	-
MMB6	Brettenham	pit	Early Saxon	2847	26	12	29	-
MPD1	Didcot	pit	Early Saxon	7/04	2	-	-	-
MPD2	Didcot	pit	Early Saxon	17/4	3	-	-	-
MPD3	Didcot	ditch	Early Saxon	24/11	4	-	-	-
MPD4	Didcot	ditch	Early Saxon	25/04	5	-	-	-
NW1	Neptune Wood	pit	Intermediate	25426	25001	40	50	1.3
NW2	Neptune Wood	pit	Intermediate	25427	25003	40	1	-
OUT1	Outwell	other	Early Saxon	O31	1	10	-	-
PL1	Pennyland	pit	Intermediate	SFB1	62	3.6	11	-
PL2	Pennyland	sfb	Intermediate	SFB1	242	5.4	7	-
PL3	Pennyland	sfb	Mid Saxon	SFB2	48	3.6	111	30.8
PL4	Pennyland	sfb	Mid Saxon	SFB2	255	1.8	8	-

code	site	feature	chronology	context	sample	volume (litres)	abundance	average density
PL5	Pennyland	sfb	Mid Saxon	SFB2	256	1.8	25	-
PL6	Pennyland	sfb	Mid Saxon	SFB2	257	3.6	98	27.2
PL7	Pennyland	sfb	Mid Saxon	SFB2	259	6	29	-
PL8	Pennyland	sfb	Mid Saxon	SFB3	699	5.4	5	-
PL9	Pennyland	sfb	Mid Saxon	SFB4	743	3.6	15	-
PL10	Pennyland	sfb	Intermediate	SFB5	689	3.6	1	-
PL11	Pennyland	sfb	Intermediate	SFB6	695	3.6	42	11.7
PL12	Pennyland	sfb	Early Saxon	SFB7	223	1.8	4	-
PL13	Pennyland	sfb	Mid Saxon	SFB8	693	1.8	20	-
PL14	Pennyland	sfb	Early Saxon	SFB9	658	1.8	11	-
PL15	Pennyland	sfb	Intermediate	SFB11	744	3.6	7	-
PL16	Pennyland	sfb	Early Saxon	SFB12	798	3.6	8	-
PL17	Pennyland	sfb	Intermediate	SFB13	807	3.6	3	-
PL18	Pennyland	other	Mid Saxon	-	737	1.8	6	-
PL19	Pennyland	other	Mid Saxon	-	792	3.6	4	-
PL20	Pennyland	other	Mid Saxon	-	793	1.8	4	-
PL21	Pennyland	pit	Early Saxon	-	463	5.4	25	-
PL22	Pennyland	pit	Intermediate	-	558	1.8	2	-
PL23	Pennyland	pit	Intermediate	-	801	1	4	-
PL24	Pennyland	ditch	Intermediate	-	630	1.8	8	-
PL25	Pennyland	ditch	Intermediate	-	959	1.8	6	-
PT1	Pitstone	pit	Early Saxon	G43	2	20	20	-
PT2	Pitstone	sfb	Early Saxon	G28	3	10	4	-
PT3	Pitstone	sfb	Early Saxon	G33	14	10	13	-
RFT2-3	Redcastle Furze	sfb	Intermediate	844	1 & 2	45	22	-
RFT4-5	Redcastle Furze	sfb	Intermediate	795	3 & 4	45	41	0.9
RFT6-7	Redcastle Furze	sfb	Intermediate	898	5 & 6	52	231	4.4
RFT8-9	Redcastle Furze	sfb	Intermediate	810	7 & 8	43.5	130	3.0
RFT10	Redcastle Furze	pit	Intermediate	1282	9	12	14	-
RFT11	Redcastle Furze	sfb	Intermediate	1298	10	18	7	-

code	site	feature	chronology	context	sample	volume (litres)	abundance	average density
RFT12	Redcastle Furze	sfb	Intermediate	1667	14	72	86	1.2
RFT13	Redcastle Furze	sfb	Intermediate	1598	11	102	54	0.5
RFT14	Redcastle Furze	pit	Intermediate	1817	21	13.5	2	-
RFT15	Redcastle Furze	sfb	Intermediate	1936	23	18	29	-
RFT16	Redcastle Furze	sfb	Intermediate	1964	24	25	17	-
RFT17	Redcastle Furze	sfb	Intermediate	2124	27	18	10	-
ROS1	Rosemary Lane	ditch	Mid Saxon	1281	100	6	471	78.5
ROS2	Rosemary Lane	ditch	Mid Saxon	1289	101	16	498	31.1
ROS3	Rosemary Lane	ditch	Mid Saxon	1291	102	14	39	2.8
ROS4	Rosemary Lane	ditch	Mid Saxon	1411	110	15	182	12.1
RY1	Rycote	ditch	Generic	1111	8	20	67	3.4
SB1	Hintlesham	ditch	Mid Saxon	60	3	-	-	-
SC1	Sutton Courtenay	pit	Intermediate	367	8	20	7	-
SC2	Sutton Courtenay	pit	Intermediate	318	6	16	14	-
SC3	Sutton Courtenay	pit	Intermediate	454	25	40	4	-
SC4	Sutton Courtenay	pit	Intermediate	404	22	40	9	-
SH1	Lechlade	sfb	Intermediate	152	-	30	13	-
SH2	Lechlade	sfb	Intermediate	154	-	30	128	4.3
SH3	Lechlade	sfb	Intermediate	242	-	30	15	-
SH4	Lechlade	sfb	Intermediate	243	-	30	575	19.2
SHB1	Benson	sfb	Intermediate	364	81	20	9	-
SHB2	Benson	pit	Intermediate	399	87	10	9	-
SHF9	Slough House Farm	well	Intermediate	196	111	-	1	-
SHF10	Slough House Farm	well	Intermediate	235	114	-	2	-
SHF11	Slough House Farm	well	Intermediate	295	115	-	6	-
SHF12	Slough House Farm	well	Intermediate	318	117	-	1	-
SMB1	Brandon (BRD 071)	pit	Early Saxon	239	-	11	1239	112.6
SMB3	Brandon	other	Mid Saxon	0055	-	-	123	-
SMB4	Brandon	other	Mid Saxon	0062	-	-	32	-
SMB5	Brandon	other	Mid Saxon	0074	-	-	15	-

code	site	feature	chronology	context	sample	volume (litres)	abundance	average density
SMB6	Brandon	other	Mid Saxon	0110	-	-	3	-
SMB7	Brandon	other	Mid Saxon	0147	-	-	31	-
SMB8	Brandon	other	Mid Saxon	0201	-	-	19	-
SMB9	Brandon	other	Mid Saxon	0215	-	-	1	-
SMB10	Brandon	other	Mid Saxon	0298	-	-	29	-
SMB11	Brandon	ditch	Mid Saxon	0539	-	1.5	1515	1010.0
SMB12	Brandon	other	Mid Saxon	1035	-	1.8	159	88.3
SMB13	Brandon	other	Mid Saxon	5038	-	-	817	-
SMB14	Brandon	other	Mid Saxon	5742	-	-	16	-
SMB15	Brandon	other	Mid Saxon	5755	-	-	328	-
SMB16	Brandon	other	Mid Saxon	5764	-	-	4706	-
SMB17	Brandon	other	Mid Saxon	6232	-	-	125	-
SMB18	Brandon	other	Mid Saxon	6457	-	-	19	-
SMB23	Brandon	other	Mid Saxon	91/59, 4293	-	-	1	-
SMB24	Brandon	other	Mid Saxon	91/59, 4297	-	-	7	-
SMB25	Brandon	other	Mid Saxon	91/59, 4301	-	-	1	-
SMB26	Brandon	other	Mid Saxon	91/60, 4333	-	-	1	-
SMB27	Brandon	other	Mid Saxon	91/60, 4345	-	-	1	-
SMB28	Brandon	other	Mid Saxon	91/61, 4681	-	-	2	-
SMB29	Brandon	other	Mid Saxon	91/61, 4683	-	-	2	-
SMB30	Brandon	other	Mid Saxon	91/61, 4684	-	-	5	-
SMB31	Brandon	other	Mid Saxon	90/61, 4966	-	-	4	-
SMB32	Brandon	other	Mid Saxon	90/61, 4967	-	-	23	-
SMB33	Brandon	other	Mid Saxon	90/61, 4969	-	-	5	-
SMB34	Brandon	other	Mid Saxon	90/61, 4970	-	-	5	-
SMB35	Brandon	other	Mid Saxon	89/61, 5059	-	-	2	-
SMB36	Brandon	other	Mid Saxon	89/61, 5062	-	-	1	-
SMB37	Brandon	other	Mid Saxon	89/61, 5063	-	-	-	-
SMB38	Brandon	other	Mid Saxon	89/61, 5064	-	-	7	-
SMB39	Brandon	other	Mid Saxon	89/61, 5065	-	-	3	-

code	site	feature	chronology	context	sample	volume (litres)	abundance	average density
SMB40	Brandon	other	Mid Saxon	88/61, 5224	-	-	1	-
SMB41	Brandon	other	Mid Saxon	88/61, 5233	-	-	4	-
SMB42	Brandon	other	Mid Saxon	88/61, 5234	-	-	5	-
SMB43	Brandon	other	Mid Saxon	89/62, 5356	-	-	4	-
SMB44	Brandon	other	Mid Saxon	89/62, 5358	-	-	15	-
SMB45	Brandon	other	Mid Saxon	89/62, 5362	-	-	3	-
SMB46	Brandon	other	Mid Saxon	89/62, 5364	-	-	6	-
SMB47	Brandon	other	Mid Saxon	89/62, 5365	-	-	3	-
SMB48	Brandon	other	Mid Saxon	90/62, 5447	-	-	5	-
SMB49	Brandon	other	Mid Saxon	90/62, 5375	-	-	2	-
SMB50	Brandon	other	Mid Saxon	90/62, 5382	-	-	1	-
SMB51	Brandon	other	Mid Saxon	91/62, 5044	-	-	-	-
SMB52	Brandon	other	Mid Saxon	91/62, 5045	-	-	4	-
SMB53	Brandon	other	Mid Saxon	91/62, 5046	-	-	1	-
SMB54	Brandon	other	Mid Saxon	91/62, 5050	-	-	1	-
SMB55	Brandon	other	Mid Saxon	90/63, 5784	-	-	4	-
SMB56	Brandon	other	Mid Saxon	90/63, 5787	-	-	1	-
SPH1	Spong Hill	sfb	Early Saxon	128-3	-	36	3	-
SPH2	Spong Hill	sfb	Early Saxon	3396	-	24	3	-
SPH3	Spong Hill	sfb	Early Saxon	3482	-	12	12	-
SPH4	Spong Hill	sfb	Early Saxon	3485	-	12	3	-
SPH5	Spong Hill	pit	Early Saxon	1621	-	12	8	-
SR1	Spring Road	sfb	Early Saxon	2479	19	40	38	1.0
SR2	Spring Road	sfb	Early Saxon	2672	22	40	54	1.4
SR3	Spring Road	sfb	Early Saxon	2673	23	40	44	1.1
SR4	Spring Road	sfb	Early Saxon	2686	34	40	57	1.4
SSA1	Stansted Airport	other	Early Saxon	2330	5	11	3	-
SSA2	Stansted Airport	pit	Early Saxon	2104	38	6	6	-
STG1	Stonea Grange	other	Early Saxon	multiple	multiple	295	-	-
TMB1	Two Mile Bottom	other	Early Saxon	multiple	multiple	-	-	-

code	site	feature	chronology	context	sample	volume (litres)	abundance	average density
TP1	Taplow Court	ditch	Mid Saxon	460	579	40	484	12.1
TP2	Taplow Court	ditch	Mid Saxon	3004	303	40	-	-
TSC1	Terrington St Clement	pit	Mid Saxon	86	1	7	33	4.7
TSC2	Terrington St Clement	pit	Mid Saxon	8	2	7	148	21.1
TSC3	Terrington St Clement	ditch	Mid Saxon	75	4	7	2	-
TSC4	Terrington St Clement	ditch	Mid Saxon	26	5	7	162	23.1
TSC5	Terrington St Clement	ditch	Mid Saxon	117	6	7	26	-
TSC6	Terrington St Clement	ditch	Mid Saxon	76	2	7	104	14.9
TSC7	Terrington St Clement	ditch	Mid Saxon	106	3	7	106	15.1
W1	Worton	posthole	Mid Saxon	multiple	multiple	30	-	-
W2	Worton	other	Mid Saxon	multiple	multiple	50	-	-
WAL1	Walton Lodge	pit	Intermediate	354	2008	-	237	-
WE1	Bletchley	ditch	Mid Saxon	143	11	30	31	1.0
WE2	Bletchley	ditch	Mid Saxon	64	7	29	20	-
WE3	Bletchley	pit	Mid Saxon	90	8	28	29	-
WE4	Bletchley	ditch	Mid Saxon	58	9	29.5	7	-
WE5	Bletchley	pit	Mid Saxon	91	1	3	69	23.0
WE6	Bletchley	sfb	Mid Saxon	193	4	25	6	-
WE7	Bletchley	ditch	Mid Saxon	48	6	29	26	-
WF1	Wickhams Field	pit	Mid Saxon	536, 542	multiple	-	-	-
WFC1	Consortium Site	pit	Mid Saxon	742	8	20	115	5.8
WFC3	Consortium Site	ditch	Mid Saxon	834	11	20	44	2.2
WFC4	Consortium Site	pit	Mid Saxon	902	16	20	43	2.2
WFC5	Consortium Site	well	Mid Saxon	922	17	20	56	2.8
WFC6	Consortium Site	well	Mid Saxon	919	18	20	379	19.0
WFC7	Consortium Site	well	Mid Saxon	950	19	20	20	-
WFC8	Consortium Site	well	Mid Saxon	970	22	20	7	-
WFC9	Consortium Site	well	Mid Saxon	969	23	20	6	-
WFC10	Consortium Site	well	Mid Saxon	971	24	20	24	-
WFC11	Consortium Site	ditch	Mid Saxon	1183	27	20	96	4.8

code	site	feature	chronology	context	sample	volume (litres)	abundance	average density
WFC12	Consortium Site	pit	Mid Saxon	1294	29	20	82	4.1
WFC13	Consortium Site	pit	Mid Saxon	296	43	20	17	-
WFC14	Consortium Site	pit	Mid Saxon	1734	45	20	95	4.8
WFC15	Consortium Site	ditch	Mid Saxon	1756	47	40	51	1.3
WFC16	Consortium Site	pit	Mid Saxon	1900	55	20	3	-
WFR1	Ashwell Site	pit	Mid Saxon	1834	16	8	434	54.3
WFR2	Ashwell Site	ditch	Mid Saxon	2223	32	8	10	-
WFR3	Ashwell Site	pit	Mid Saxon	2519	66	12	11	-
WFR4	Ashwell Site	posthole	Mid Saxon	2642	71	4	137	34.3
WFR5	Ashwell Site	pit	Mid Saxon	7162	165	8	315	39.4
WFR6	Ashwell Site	pit	Mid Saxon	7163	162	5	17	-
WG1	Wavendon Gate	ditch	Early Saxon	1102	85	15	15	-
WG2	Wavendon Gate	ditch	Early Saxon	1135	88	15	8	-
WG3	Wavendon Gate	pit	Early Saxon	1158	100	15	3	-
WG4	Wavendon Gate	pit	Early Saxon	1157	101	15	6	-
WHR1	Whitehouse Road	other	Mid Saxon	526	-	7	-	-
WHR2	Whitehouse Road	other	Mid Saxon	559	-	7.5	-	-
WHR3	Whitehouse Road	other	Mid Saxon	1692	-	7	-	-
WHR5	Whitehouse Road	other	Mid Saxon	1725	-	7	-	-
WHR6	Whitehouse Road	other	Mid Saxon	1735	-	7	-	-
WHR7	Whitehouse Road	other	Mid Saxon	1741	-	7	-	-
WIL1	Wilton	hearth/oven	Intermediate	596	146	5	-	-
WIL2	Wilton	hearth/oven	Intermediate	595	147	8	-	-
WIL3	Wilton	hearth/oven	Intermediate	594	148	0.5	-	-
WIL4	Wilton	hearth/oven	Intermediate	597	149	0.5	-	-
WIL5	Wilton	hearth/oven	Intermediate	609	155	3	-	-
WIL6	Wilton	hearth/oven	Intermediate	610	156	10	-	-
WIL7	Wilton	hearth/oven	Intermediate	612	158	0.5	-	-
WIL8	Wilton	hearth/oven	Intermediate	614	160	2	-	-
WIL9	Wilton	hearth/oven	Intermediate	615	161	3	-	-

code	site	feature	chronology	context	sample	volume (litres)	abundance	average density
WIL10	Wilton	sfb	Intermediate	550	124	20	-	-
WIL11	Wilton	sfb	Intermediate	550	139	18	-	-
WIL12	Wilton	sfb	Intermediate	550	181	10	-	-
WIL13	Wilton	posthole	Intermediate	653	182	3	-	-
WIL14	Wilton	posthole	Intermediate	655	183	6	-	-
WIL15	Wilton	pit	Intermediate	586	140	4	-	-
WIL16	Wilton	pit	Intermediate	622	168	20	-	-
WIL17	Wilton	posthole	Intermediate	690	197	10	-	-
WIL18	Wilton	posthole	Intermediate	690	198	10	-	-
WIL19	Wilton	pit	Intermediate	392	88	10	-	-
WIL20	Wilton	pit	Intermediate	342	81	1	-	-
WKB1	Wicken Bonhunt	ditch	Intermediate	F605	B	180	249	1.4
WKB2	Wicken Bonhunt	ditch	Mid Saxon	F914	C	180	1	-
WLP1	Walpole St Andrew	pit	Mid Saxon	29	1	7	27	-
WLP2	Walpole St Andrew	ditch	Mid Saxon	129	2	7	128	18.3
WLP3	Walpole St Andrew	other	Mid Saxon	188	3	7	102	14.6
WLP4	Walpole St Andrew	pit	Mid Saxon	216	5	7	86	12.3
WLP5	Walpole St Andrew	pit	Mid Saxon	300	7	7	116	16.6
WLP6	Walpole St Andrew	pit	Mid Saxon	239	8	8	120	15.0
WLP7	Walpole St Andrew	ditch	Mid Saxon	199	9	10.5	265	25.2
WLP8	Walpole St Andrew	ditch	Mid Saxon	231	10	9	90	10.0
WLP9	Walpole St Andrew	ditch	Mid Saxon	40	11	9	292	32.4
WLP10	Walpole St Andrew	ditch	Mid Saxon	17	12	7	117	16.7
WLP11	Walpole St Andrew	pit	Mid Saxon	106	13	7	12	-
WO1	Walton Orchard	other	Intermediate	270	-	10	15	-
WO2	Walton Orchard	other	Intermediate	650	-	40	15	-
WO3	Walton Orchard	other	Intermediate	652	-	10	5	-
WO4	Walton Orchard	other	Intermediate	777	-	30	9	-
WO5	Walton Orchard	other	Intermediate	866	-	20	10	-
WO6	Walton Orchard	other	Intermediate	987	-	10	5	-

code	site	feature	chronology	context	sample	volume (litres)	abundance	average density
WO7	Walton Orchard	other	Intermediate	1374	-	10	6	-
WO8	Walton Orchard	other	Intermediate	1693	-	10	9	-
WO9	Walton Orchard	other	Intermediate	1878	-	5	4	-
WO10	Walton Orchard	other	Intermediate	2076	-	10	9	-
WRS1	Walton Road Stores	sfb	Generic	2171	3002	-	126	-
WRS2	Walton Road Stores	sfb	Generic	2178	3003	-	70	-
WRS3	Walton Road Stores	sfb	Generic	2389	3014	-	31	-
WRS4	Walton Road Stores	sfb	Generic	2475	3020	-	750	-
WRS5	Walton Road Stores	sfb	Generic	2480	3019	-	240	-
WRS6	Walton Road Stores	sfb	Generic	367	3502	-	81	-
WRS7	Walton Road Stores	pit	Generic	271	3505	-	87	-
WRS8	Walton Road Stores	sfb	Generic	577	3519	-	126	-
WRS9	Walton Road Stores	sfb	Generic	482	3521	-	95	-
WSA1	Walton Street	posthole	Intermediate	3010	1	30	17	-
WSA2	Walton Street	ditch	Intermediate	3014	3	40	1	-
WSA3	Walton Street	ditch	Intermediate	3016	4	40	9	-
WST1	West Stow	sfb	Early Saxon	63	1	4108.7	2085	0.5
WST2	West Stow	pit	Generic	59	3	-	496	-
WT1	Wolverton Turn	ditch	Mid Saxon	1	-	-	3	-
WT2	Wolverton Turn	ditch	Mid Saxon	1	-	-	1	-
WT3	Wolverton Turn	ditch	Mid Saxon	1	-	-	31	-
WT4	Wolverton Turn	ditch	Mid Saxon	1	-	-	111	-
WT5	Wolverton Turn	ditch	Mid Saxon	1	-	-	51	-
WT6	Wolverton Turn	ditch	Mid Saxon	5	-	-	17	-
WT7	Wolverton Turn	ditch	Mid Saxon	5	-	-	12	-
WT8	Wolverton Turn	other	Mid Saxon	23	-	-	7	-
WT9	Wolverton Turn	ditch	Mid Saxon	8	-	-	27	-
WT10	Wolverton Turn	ditch	Mid Saxon	8	-	-	43	-
WT11	Wolverton Turn	ditch	Mid Saxon	9	-	-	43	-
WT12	Wolverton Turn	ditch	Mid Saxon	6	-	-	7	-

code	site	feature	chronology	context	sample	volume (litres)	abundance	average density
WT13	Wolverton Turn	ditch	Mid Saxon	7	-	-	5	-
WT14	Wolverton Turn	ditch	Mid Saxon	7	-	-	39	-
WT15	Wolverton Turn	ditch	Mid Saxon	1(429)	-	-	6	-
WT16	Wolverton Turn	ditch	Mid Saxon	1(430)	-	-	5	-
WT17	Wolverton Turn	ditch	Mid Saxon	1(430)	-	-	4	-
WT18	Wolverton Turn	hearth/oven	Mid Saxon	418, 489	-	-	2	-
WTN1	Witton	sfb	Early Saxon	Structure D	-	420	106	0.3
WTN2	Witton	sfb	Early Saxon	Structure C	-	1113	362	0.3
WV1	Walton Vicarage	pit	Generic	P5	632 (upper)	-	-	-
WV2	Walton Vicarage	pit	Generic	P5	632 (lower)	-	-	-
WV3	Walton Vicarage	pit	Generic	P5	634	-	-	-
WWI1	Ingleborough	ditch	Mid Saxon	85	6	7	4	-
WWI2	Ingleborough	ditch	Mid Saxon	86	7	7	7	-
WWI3	Ingleborough	ditch	Mid Saxon	88	8	7	-	-
WWI4	Ingleborough	other	Mid Saxon	80	9	7.5	232	30.9
WWI5	Ingleborough	ditch	Mid Saxon	95	11	7	18	-
WWI6	Ingleborough	pit	Mid Saxon	109	14	7	4	-
WWI7	Ingleborough	ditch	Mid Saxon	115	17	10.5	279	26.6
WWI8	Ingleborough	other	Mid Saxon	102	18	7	11	-
WWI9	Ingleborough	other	Mid Saxon	186	19	7	9	-
WYM1	Wymondham	hearth/oven	Generic	64	1	10	-	-
Y1	Yarnton	other	Early Saxon	354	70 & 72	20	30	1.5
Y2	Yarnton	other	Early Saxon	888	51	10	17	-
Y3	Yarnton	pit	Early Saxon	2050	52	10	44	4.4
Y4	Yarnton	sfb	Early Saxon	2652	82 & 86	20	75	3.8
Y5	Yarnton	sfb	Early Saxon	2689	77	10	65	6.5
Y6	Yarnton	sfb	Intermediate	2551	63	10	15	-
Y7	Yarnton	sfb	Intermediate	2552	66	10	21	-
Y8	Yarnton	sfb	Intermediate	2556	64	10	24	-
Y9	Yarnton	sfb	Intermediate	2561	65	10	8	-

code	site	feature	chronology	context	sample	volume (litres)	abundance	average density
Y10	Yarnton	sfb	Intermediate	2563	62	10	27	-
Y11	Yarnton	sfb	Intermediate	2577	67	10	11	-
Y12	Yarnton	sfb	Intermediate	3004	275	10	70	7.0
Y13	Yarnton	sfb	Intermediate	3036	280 & 281	20	16	-
Y14	Yarnton	ditch	Intermediate	3045	264	10	22	-
Y15	Yarnton	posthole	Intermediate	3192	798	5	10	-
Y16	Yarnton	posthole	Intermediate	3209	796	5	10	-
Y17	Yarnton	posthole	Intermediate	3221	794	5	10	-
Y18	Yarnton	posthole	Intermediate	3256	799	5	6	-
Y19	Yarnton	ditch	Intermediate	3279	271 & 267	20	36	1.8
Y20	Yarnton	ditch	Intermediate	3287	multiple	40	13	-
Y21	Yarnton	hearth/oven	Intermediate	3305	241	10	8	-
Y22	Yarnton	ditch	Intermediate	3572	286 & 287	20	78	3.9
Y23	Yarnton	posthole	Intermediate	3891	256 & 257	10	1	-
Y24	Yarnton	pit	Intermediate	3928	290 & 291	20	31	1.6
Y25	Yarnton	posthole	Mid Saxon	2616	61	5	390	78.0
Y26	Yarnton	ditch	Mid Saxon	3031	220 & 221	20	141	7.1
Y27	Yarnton	ditch	Mid Saxon	3035	218	10	24	-
Y28	Yarnton	ditch	Mid Saxon	3039	224 & 225	20	19	-
Y29	Yarnton	pit	Mid Saxon	3040	226 & 227	20	21	-
Y30	Yarnton	pit	Mid Saxon	3043	multiple	40	957	23.9
Y31	Yarnton	ditch	Mid Saxon	3057	222 & 223	20	400	20.0
Y33	Yarnton	ditch	Mid Saxon	3176	770 & 769	20	9	-
Y34-5	Yarnton	other	Mid Saxon	3314	791 & 812	20	739	37.0
Y36	Yarnton	posthole	Mid Saxon	3317	707	5	45	9.0
Y37	Yarnton	posthole	Mid Saxon	3392	703	5	6	-
Y38	Yarnton	ditch	Mid Saxon	3429	815	10	181	18.1
Y39	Yarnton	ditch	Mid Saxon	3431	811	10	706	70.6
Y40	Yarnton	posthole	Mid Saxon	3436	702	5	7	-
Y41	Yarnton	ditch	Mid Saxon	3512	292 & 293	20	26	-

code	site	feature	chronology	context	sample	volume (litres)	abundance	average density
Y42	Yarnton	posthole	Mid Saxon	3527	706	5	5	-
Y43	Yarnton	posthole	Mid Saxon	3538	705	5	14	-
Y44	Yarnton	ditch	Mid Saxon	3629	789 & 790	20	67	3.4
Y45	Yarnton	ditch	Mid Saxon	3641	787 & 788	20	18	-
Y46	Yarnton	ditch	Mid Saxon	3646	808 & 809	20	6	-
Y47	Yarnton	pit	Mid Saxon	3693	246 & 248	20	20153	1007.7
Y48	Yarnton	posthole	Mid Saxon	3738	701	5	16	-
Y49	Yarnton	other	Mid Saxon	5042	754	10	303	30.3

Appendix 5: Inventory of Plant Taxa

Percentages are derived from totals of 111 assemblages and 736 samples.

Classes are defined as follows: (A) probable crops, (B) possible crops, (C) possible weeds, (D) non-arable, and (E) indeterminate.

Crops: cereals

taxon	class	% assemblages	% samples
All cereals	A/B	99.1	95.8
Avena sativa L.	A	4.5	1.0
Avena L. (indet.)	B	64.0	34.6
Cereal indet.	A	90.1	83.7
Hordeum L.	A	92.8	66.0
Secale cereale L.	A	48.6	26.8
Triticum dicoccum Schübl.	A	11.7	3.1
Triticum dicoccum Schübl. / *spelta* L.	A	27.0	9.4
Triticum L. (indet.)	A	74.8	43.3
Triticum L. (free-threshing)	A	64.0	40.5
Triticum spelta L.	A	29.7	8.8

Crops: legumes

taxon	class	% assemblages	% samples
All legumes	A/B	69.4	36.5
Large legume indet.	B	58.6	30.4
Lathyrus L.	B	2.7	0.5
Lens culinaris (Medik.)	A	1.8	0.3
Pisum sativum L.	A	12.6	4.8
Vicia L.	B	8.1	2.6
Vicia faba L.	A	17.1	4.5
Vicia faba L./*Pisum sativum* L.	A	1.8	0.3
Vicia sativa L.	B	3.6	0.8

Crops: others

taxon	class	% assemblages	% samples
Cannabis sativa L.	A	0.9	0.1
Humulus lupulus L.	A	0.9	0.1
Linum usitatissimum L.	A	15.3	4.2
Papaver somniferum L.	A	2.7	0.5
Vitis vinifera L.	A	1.8	0.3

Possible weeds (Class C)

taxon	% assemblages	% samples
Achillea L. / *Anthemis* L. / *Artemisia* L. / *Tripleurospermum* (Sch. Bip.)	1.8	0.3
Aethusa cynapium L.	1.8	0.5
Agrimonia eupatoria L.	0.9	0.1
Agrostemma githago L.	18.0	6.1
Agrostis capillaris L.	0.9	0.1
Agrostis L./*Poa* L.	0.9	0.1
Alchemilla L.	0.9	0.1
Alismataceae	2.7	0.4
Allium L.	0.9	0.1
Amaranthaceae	34.2	14.4
Anagallis arvensis L.	2.7	1.1
Anisantha sterilis (L.) Nevski	4.5	1.0
Anthemis arvensis L.	0.9	0.1
Anthemis cotula L.	36.0	14.0
Aphanes arvensis L. + *Potentilla* L.	0.9	0.3
Aphanes arvensis L./*australis* Rydb.	2.7	0.4
Apiaceae	11.7	2.2
Apium L.	1.8	0.5
Arctium L.	0.9	0.1
Arenaria serpyllifolia L.	0.9	0.1
Arrhenatherum (P. Beauv.)/*Avena* L.	0.9	1.0
Arrhenatherum elatius (L.) P. Beauv ex. J. & C. Presl	5.4	1.5
Asteraceae	19.8	5.6
Atriplex patula L./*prostrata* (Boucher ex. DC.)	34.2	11.4
Avena fatua L.	4.5	0.8
Avena L./*Bromus* L.	9.9	2.7
Avena sterilis L.	0.9	0.1
Ballota nigra L.	0.9	0.1
Beta vulgaris L.	0.9	0.3
Brachypodium P. Beauv.	0.9	0.1

Brassica L.	7.2	1.9
Brassica L./*Sinapis* L.	12.6	3.0
Brassica nigra (L.) W.D.J. Koch	1.8	0.5
Brassica rapa ssp *campestris* (L.) A.R. Clapham	1.8	0.3
Brassicaceae	8.1	1.5
Bromus hordeaceus L./*secalinus* L.	40.5	17.7
Bupleurum rotundifolium L.	5.4	0.7
Caltha palustris L.	0.9	0.1
Camelina sativa (L.) Crantz	3.6	1.1
Carduus L./*Cirsium* Mill.	1.8	0.3
Carex distans L./*sylvatica* (Huds.)/*laevigata* (Sm.)	1.8	0.5
Carex flava L.	0.9	0.1
Carex L.	35.1	12.9
Carex panicea L.	0.9	0.1
Caryophyllaceae	9.9	1.9
Centaurea cyanus L.	1.8	0.3
Centaurea L.	9.0	1.6
Centaurea nigra L.	1.8	0.3
Cerastium L.	1.8	0.3
Cerastium L./*Stellaria media* (L.) Vill.	3.6	0.7
Chelidonium L.	0.9	0.1
Chenopodium album L.	38.7	15.6
Chenopodium ficifolium (Sm.)	6.3	1.4
Chenopodium hybridum L.	1.8	0.3
Chenopodium L.	21.6	7.5
Chenopodium L./*Atriplex* L.	0.9	0.1
Chenopodium polyspermum L.	4.5	0.7
Chenopodium rubrum L./*glaucum* L.	1.8	0.3
Cladium mariscus (L.) Pohl	8.1	2.0
Conium L./*Pimpinella* L.	0.9	0.1
Conium maculatum L.	0.9	0.0
Conopodium majus (Gouan) Loret	0.9	0.3
Convolvulus arvensis L.	0.9	0.3
Cuscuta L.	0.9	0.1
Cyperaceae	12.6	3.9
Cyperaceae/Polygonaceae	0.9	0.1
Danthonia decumbens (L.) DC.	0.9	0.1
Daucus carota L.	0.9	0.1
Deschampsia cespitosa (L.) P. Beauv.	1.8	0.3
Eleocharis palustris (L.) Roem. & Schult./*uniglumis* (Link) Schult.	35.1	14.3
Epilobium L.	2.7	1.2

Eriophorum L.	0.9	0.1
Euphorbia helioscopia L.	0.9	0.1
Euphrasia L./*Odontites vernus* (Bellardi) Dumort.	9.0	1.8
Fabaceae	26.1	10.6
Fallopia convolvulus (L.) Á. Löve	38.7	14.7
Festuca L.	4.5	1.1
Fumaria officinalis L.	1.8	0.3
Galeopsis tetrahit L.	4.5	0.8
Galium album (Mill.)/*palustre* L./*saxatile* L./*spurium* L./*verum* L.	2.7	0.7
Galium album Mill.	0.9	0.1
Galium aparine L.	34.2	11.5
Galium L.	9.9	2.9
Galium palustre L.	1.8	0.3
Geranium L.	0.9	0.1
Glaux maritima L.	0.9	0.3
Glebionis segetum (L.) Fourr.	0.9	0.1
Helminthotheca echioides (L.) Holub	0.9	0.3
Hordeum murinum L.	0.9	0.1
Hyoscyamus niger L.	8.1	1.8
Iris pseudacorus L.	2.7	0.4
Juncus L.	8.1	2.9
Lamiaceae	5.4	1.4
Lamium L.	0.9	0.1
Lapsana communis L.	3.6	1.6
Lathyrus aphaca L.	0.9	0.1
Lathyrus nissolia L.	2.7	0.4
Lathyrus sylvestris L.	0.9	0.1
Leontodon L.	0.9	0.1
Lepidium campestre (L.) W.T. Aiton	0.9	0.3
Leucanthemum vulgare (Lam.)	1.8	0.3
Linum catharticum L.	0.9	0.1
Linum L.	2.7	0.4
Lithospermum arvense L.	7.2	1.4
Lolium L.	3.6	1.6
Lolium L./*Festuca* L.	6.3	1.4
Lolium perenne L.	2.7	0.4
Lolium temulentum L.	3.6	0.5
Lysimachia L./*Anagallis* L.	0.9	0.0
Lythrum salicaria L.	0.9	0.1
Malva sylvestris L.	24.3	8.2
Medicago L.	3.6	0.8

Medicago L./*Melilotus* Mill./*Trifolium* L.	2.7	0.4
Medicago L./*Trifolium* L.	13.5	3.3
Medicago L./*Trifolium* L./*Lotus* L.	8.1	2.9
Medicago lupulina L.	9.0	2.6
Medicago lupulina L./*Trifolium* L.	0.9	0.3
Mentha arvensis L./*aquatica* L.	6.3	1.0
Montia fontana L.	5.4	0.7
Myosotis arvensis (L.) Hill	1.8	0.3
Odontites vernus (Bellardi) Dumort.	10.8	3.0
Onopordum acanthium L.	1.8	0.5
Oxalis acetosella L.	0.9	0.1
Papaver argemone L.	0.9	0.1
Papaver L.	2.7	0.5
Papaver rhoeas L./*dubium* L.	1.8	0.1
Pastinaca sativa L.	0.9	0.1
Persicaria (Mill.)	0.9	0.7
Persicaria lapathifolia (L.) Delarbre	3.6	1.1
Persicaria maculosa (Gray)	8.1	2.2
Persicaria maculosa (Gray)/*lapathifolia* (L.) Delarbre	6.3	1.5
Persicaria minor (Huds.) Opiz	1.8	0.3
Phleum bertolonii DC.	0.9	0.1
Phleum L.	6.3	1.6
Phleum L./*Poa* L.	3.6	1.1
Phleum pratense L.	1.8	0.3
Phragmites australis (Cav.) Trin. Ex Steud.	1.8	1.4
Picris hieracioides L.	0.9	0.1
Plantago L./*Sherardia* L.	0.9	0.1
Plantago lanceolata L.	25.2	7.6
Plantago major L.	6.3	2.2
Plantago maritima L.	1.8	0.8
Plumbaginaceae	0.9	0.1
Poa annua L.	5.4	1.2
Poa L.	6.3	1.6
Poa pratensis L./*trivialis* L.	1.8	0.3
Poaceae	66.7	28.9
Polygala L.	0.9	0.0
Polygonaceae	17.1	6.7
Polygonum arenastrum Boreau	0.9	0.1
Polygonum aviculare L.	32.4	8.6
Polygonum aviculare L./*Persicaria* (Mill.)	0.9	0.1
Polygonum L.	9.9	1.8

Polygonum L./*Fallopia* (Adans.)	0.9	0.1
Potentilla L.	2.7	0.4
Primulaceae	0.9	0.1
Prunella vulgaris L.	6.3	1.1
Pteridium aquilinum (L.) Kuhn	4.5	1.2
Ranunculus acris L./*bulbosus* L./*repens* L.	12.6	2.6
Ranunculus arvensis L./*parviflorus* L.	0.9	0.1
Ranunculus flammula L.	4.5	1.0
Ranunculus L.	9.9	1.8
Ranunculus repens L.	6.3	1.0
Ranunculus sceleratus L.	0.9	0.1
Ranunculus subg. *ranunculus* L.	2.7	0.8
Raphanus raphanistrum L.	12.6	2.7
Reseda luteola L.	3.6	0.5
Rhinanthus minor L.	1.8	0.4
Rubiaceae	0.9	0.1
Rumex acetosa L.	1.8	0.3
Rumex acetosella L.	19.8	5.0
Rumex conglomeratus (Murray)/*obtusifolius* L./*sanguineus* L.	5.4	1.4
Rumex crispus L.	9.0	3.1
Rumex crispus L./*longifolius* DC.	0.9	0.1
Rumex L.	61.3	19.0
Sanguisorbia officinalis L.	0.9	0.1
Schoenus nigricans L.	2.7	1.0
Scirpus L.	3.6	1.6
Scleranthus annuus L.	3.6	0.7
Scutellaria L.	0.9	0.1
Sedum acre L.	0.9	0.1
Sherardia arvensis L.	5.4	0.8
Silene flos-cuculi (L.) Clairv.	1.8	0.3
Silene gallica L.	0.9	0.1
Silene L.	9.9	2.2
Silene latifolia (Poir.)	2.7	0.4
Silene vulgaris (Moench) Garcke	0.9	0.1
Sinapis alba L./*arvensis* L.	2.7	0.7
Solanum nigrum L.	6.3	1.4
Sonchus asper (L.) Hill	0.9	0.1
Sparganium erectum L.	3.6	0.5
Spergula arvensis L.	9.9	2.3
Stachys sylvatica L.	1.8	0.3
Stellaria graminea L.	3.6	0.7

Stellaria L.	2.7	0.5
Stellaria media (L.) Vill.	13.5	2.9
Tanacetum vulgare L.	0.9	0.1
Thlaspi arvense L.	7.2	1.1
Torilis arvensis (Huds.) Link/*japonica* (Houtt.) DC.	3.6	1.1
Trifolium arvense L./*campestre* (Shreb.)/*dubium* (Sibth.)/*repens* L.	4.5	1.0
Trifolium L.	12.6	2.6
Trifolium L./*Lotus* L.	0.9	0.1
Trifolium pratense L.	1.8	0.3
Trifolium pratense L./*repens* L.	2.7	0.7
Triglochin maritima L.	0.9	0.1
Tripleurospermum maritimum (L.) W.D.J. Koch/*inodorum* (L.) Sch. Bip.	10.8	3.0
Urtica dioica L.	6.3	1.0
Urtica L.	1.8	0.4
Urtica urens L.	9.0	1.8
Valerianella dentata (L.) Pollich	5.4	1.0
Verbascum L.	0.9	0.1
Veronica arvensis L.	1.8	0.1
Veronica hederifolia L.	2.7	0.4
Veronica L.	1.8	0.3
Vicia tetrasperma (L.) Schreb.	2.7	0.5
Viola L.	0.9	0.1

Non-arable taxa (Class D)

taxon	% assemblages	% samples
Bryophyta	1.8	0.4
Bolboschoenus maritimus (L.) Palla	3.6	2.3
Bolboschoenus maritimus (L.) Palla / *Schoenoplectus lacustris* (L.) Palla	0.9	0.0
Calluna vulgaris (L.) Hull	8.1	4.1
Corylus avellana L.	38.7	17.8
Crataegus monogyna (Jacq.)	4.5	1.1
Ericaceae	9.0	4.1
Malus sylvestris (L.) Mill. / *pumila* (Mill.)	3.6	0.7
Prunus avium (L.) L.	1.8	0.4
Prunus avium (L.) L. / *spinosa* L.	0.9	0.1
Prunus domestica L.	1.8	0.3
Prunus L.	5.4	1.1
Prunus L. / *Crataegus* L.	2.7	0.7
Prunus spinosa L.	7.2	1.2
Rhamnus cathartica L.	0.9	0.1
Rosa L.	0.9	0.1
Rubus fruticosus L.	0.9	0.5
Rubus idaeus L.	0.9	0.1
Rubus L.	5.4	1.1
Sambucus nigra L.	9.9	3.3
Schoenoplectus lacustris (L.) Palla	1.8	0.5
Sorbus L.	0.9	0.1

Indeterminate (Class E)

taxon	% assemblages	% samples
Indeterminate	70.3	41.0

Bibliography

Abrams, J. and D. Ingham 2008. *Farming on the Edge: Archaeological Evidence from the Clay Uplands to the West of Cambridge*. Bedford: Albion Archaeology.

Allen, T., N. Barton and A. Brown 1995. *Lithics and Landscape: archaeological discoveries on the Thames Water pipeline at Gatehampton Farm, Goring, Oxfordshire, 1985-92*. Oxford: Oxford University Committee for Archaeology.

Allen, T., K. Cramp, H. Lamdin-Whymark and L. Webley 2010. *Castle Hill and its Landscape; Archaeological Investigations at the Wittenhams, Oxfordshire*. Oxford: Oxford Archaeology.

Allen, T., C. Hayden and H. Lamdin-Whymark 2009. *From Bronze Age enclosure to Anglo-Saxon settlement. Archaeological excavations at Taplow hillfort, Buckinghamshire, 1999-2005*. Oxford: Oxford Archaeology.

Allen, T. and Z. Kamash 2008. *Saved from the grave: Neolithic to Saxon discoveries at Spring Road Municipal Cemetery, Abingdon, Oxfordshire, 1990-2000*. Oxford: Oxford Archaeology.

Ames, J. 2005. An Archaeological Evaluation at Browick Road, Wymondham, Norfolk. Unpublished report, Norfolk Archaeological Unit.

Andrews, P. 1995. *Excavations at Redcastle Furze, Thetford, 1988-9*. Gressenhall: Norfolk Museums Service.

Arnold, C.J. and P. Wardle 1981. Early medieval settlement patterns in England. *Medieval Archaeology* 25: 145–149.

Arthur, P., G. Fiorentino and A.M. Grasso 2012. Roads to recovery: an investigation of early medieval agrarian strategies in Byzantine Italy in and around the eighth century. *Antiquity* 86: 444–455.

Astill, G. 1997. An archaeological approach to the development of agricultural technologies in medieval England, in G. Astill and J. Langdon (eds) *Medieval farming and technology: the impact of agricultural change in northwest Europe*: 193-223. Leiden: Brill.

Atkins, R. and A. Connor 2010. *Farmers and Ironsmiths: Prehistoric, Roman and Anglo-Saxon Settlements beside Brandon Road, Thetford, Norfolk*. Bar Hill: Oxford Archaeology East.

Banham, D. 1990. The Knowledge and Uses of Food Plants in Anglo-Saxon England. Unpublished PhD dissertation, University of Cambridge.

Banham, D. 2004. *Food and Drink in Anglo-Saxon England*. Stroud: Tempus Publishing.

Banham, D. 2010. "In the Sweat of thy Brow Shalt thou eat Bread": Cereals and Cereal Production in the Anglo-Saxon Landscape, in N.J. Higham and M.J. Ryan (eds) *The Landscape Archaeology of Anglo-Saxon England*: 175-192. Woodbridge: Boydell Press.

Bateman, C., D. Enright and N. Oakey 2003. Prehistoric to Anglo-Saxon settlements to the rear of Sherborne House, Lechlade: excavations in 1997. *Transactions of the Bristol and Gloucestershire Archaeological Society* 121: 23–96.

Bates, S. and A. Lyons 2003. *The Excavation of Romano-British Pottery Kilns at Ellingham, Postwick and Two Mile Bottom, Norfolk, 1995-7*. Dereham: Norfolk Museums and Archaeological Service.

Behre, K.E. 1992. The history of rye cultivation in Europe. *Vegetation History and Archaeobotany* 1: 141–156.

Behre, K.E. 2008. Collected seeds and fruits from herbs as prehistoric food. *Vegetation History and Archaeobotany* 17: 65–73.

Blair, J. 2005. *The Church in Anglo-Saxon Society*. Oxford: Oxford University Press.

Blair, J. 2013. *The British Culture of Anglo-Saxon Settlement (H.M. Chadwick Memorial Lectures 24)*. Cambridge: Department of Anglo-Saxon, Norse and Celtic, University of Cambridge.

Blinkhorn, P. 2012. *The Ipswich ware project: ceramics, trade and society in Middle Saxon England*. London: Medieval Pottery Research Group.

Boardman, S. and G. Jones 1990. Experiments on the effects of charring on cereal plant components. *Journal of Archaeological Science* 17: 1–11.

Bogaard, A. 2004. *Neolithic Farming in Central Europe*. London: Routledge.

Bogaard, A. 2011. *Plant use and crop husbandry in an early Neolithic village: Vaihingen an der Enz, Baden-Württemberg*. Bonn: Habelt.

Bogaard, A., G. Jones and M. Charles 2005. The impact of crop processing on the reconstruction of crop sowing time and cultivation intensity from archaeobotanical weed evidence. *Vegetation History and Archaeobotany* 14: 505–509.

Bogaard, A., G. Jones, M. Charles and J.G. Hodgson 2001. On the archaeobotanical inference of crop sowing time using the FIBS method. *Journal of Archaeological Science* 28: 1171–1183.

Bonner, D. 1997. Untitled draft typescript concerning excavations at Walton, Aylesbury, on file at Buckinghamshire Historic Environment Records.

Booth, P., A. Dodd, M. Robinson and A. Smith 2007. *The Thames through Time. The Archaeology of the Gravel Terraces of the Upper and Middle Thames. The early historical period: AD 1-1000*. Oxford: Oxford Archaeology.

Booth, P., J.A. Evans and J. Hiller 2001. *Excavations in the extramural settlement of Roman Alchester, Oxfordshire, 1991*. Oxford: Oxford Archaeology.

Boulter, S. 2005. Handford Road, Ipswich (IPS 280), Archaeological Assessment Report (Volume I: Text). Unpublished report, Suffolk County Council Archaeological Service.

Boulter, S. 2008. An Assessment of the Archaeology Recorded in New Phases 5, 6, 7(a & b), 9, 11 & 12 of Flixton Park Quarry. Unpublished report, Suffolk County Council Archaeological Service.

Boulter, S. 2010. Archaeological Assessment Report. Silver Birches, Hintlesham (HNS 027). Unpublished report, Suffolk County Council Archaeological Service.

Bowen, H.C. 1961. *Ancient Fields. A tentative analysis of vanishing earthworks and landscapes*. London: British Association for the Advancement of Science.

ter Braak, C. and P. Smilauer 2002. *CANOCO Reference Manual and CanoDraw for Windows User's Guide: Software for Canonical Community Ordination (version 4.5)*. Ithaca NY/Wageningen: Microcomputer Power/Biometris.

Bradley, R. 2006. Bridging two cultures - commercial archaeology and the study of prehistoric Britain. *The Antiquaries Journal* 86: 1–13.

Brown, T. and G. Foard 1998. The Saxon landscape: a regional perspective, in P. Everson and T. Williamson (eds) *The Archaeology of Landscape*: 67-94. Manchester: Manchester University Press.

Büntgen, U., W. Tegel, K. Nicolussi, M. McCormick, D. Frank, V. Trouet, J.O. Kaplan, F. Herzig, K.U. Heussner, H. Wanner, J. Luterbacher and J. Esper 2011. 2500 Years of European Climate Variability and Human Susceptibility. *Science* 331: 578–582.

Campbell, G., L. Moffett and V. Straker 2011. *Environmental Archaeology. A Guide to the Theory and Practice of Methods, from Sampling and Recovery to Post-excavation*. 2nd ed. Swindon: English Heritage.

Campbell, G. and V. Straker 2003. Prehistoric crop husbandry and plant use in southern England: development and regionality, in K. Robson Brown (ed.) *Archaeological Sciences 1999. Proceedings of the Archaeological Sciences Conference, University of Bristol, 1999*: 14-30. Oxford: British Archaeological Reports.

Cappers, R.T.J. and R. Neef 2012. *Handbook of Plant Palaeoecology*. Groningen: Barkhuis and Groningen University Library.

Caruth, J. 1996. Ipswich, Hewlett Packard plc, Whitehouse Industrial Estate. *Proceedings of the Suffolk Institute of Archaeology and History* 38: 476–477.

Caruth, J. 2006. Consolidated Support Complex, RAF Lakenheath, ERL 116 and Family Support Complex, RAF Lakenheath ERL 139. A Report on the Archaeological Excavations, 2001-2005. Unpublished report, Suffolk County Council Archaeological Service.

Caruth, J. 2008. Eye, Hartismere High School. *Proceedings of the Suffolk Institute of Archaeology and History* 41: 518–520.

Cavers, P. and J. Harper 1964. Rumex Obtusifolius L. and R. Crispus. *Journal of Ecology* 52: 737–766.

Chambers, R. and E. McAdam 2007. *Excavations at Barrow Hills, Radley, Oxfordshire, 1983-5*. Oxford: Oxford Archaeology.

Chapman, H. 2006. *Landscape Archaeology and GIS*. Stroud: Tempus Publishing.

Charles, M., G. Jones and J.G. Hodgson 1997. FIBS in archaeobotany: Functional interpretation of weed floras in relation to husbandry practices. *Journal of Archaeological Science* 24: 1151–1161.

Clapham, A.R., T.G. Tutin and E.F. Warburg 1962. *Flora of the British Isles*. 2nd ed. Cambridge: Cambridge University Press.

Coleman, L., A. Hancocks and M. Watts 2006. *Excavations on the Wormington to Tirley Pipeline, 2000. Four sites by the Carrant Brook and River Isbourne Gloucestershire and Worcestershire*. Cirencester: Cotswold Archaeology.

Comeau, R. 2019. The practice of "in rodwallis": medieval Welsh agriculture in north Pembrokeshire, in R. Comeau and A. Seaman (eds) *Living off the Land: Agriculture in Wales c. 400 to 1600 AD*. Oxford: Windgather.

Cowie, R. 2001. English Wics: Problems with Discovery and Interpretation, in D. Hill and R. Cowie (eds) *Wics. The Early Mediaeval Trading Centres of Northern Europe*: 14-21. Sheffield: Sheffield Academic Press.

Cowie, R., L. Blackmore, A. Davis, J. Keily and K. Rielly 2012. *Lundenwic: excavations in Middle Saxon London, 1987-2000*. London: Museum of London Archaeology.

Crabtree, P.J. 2010. Agricultural innovation and socio-economic change in early medieval Europe: evidence from Britain and France. *World Archaeology* 42: 122-136.

Crabtree, P.J. 2012. *Middle Saxon animal husbandry in East Anglia*. Bury St Edmunds: Suffolk County Council Archaeological Service.

Crockett, A. 1996. Iron Age to Saxon Settlement at Wickhams Field, Near Reading, Berkshire: Excavations on the Site of the M4 Motorway Service Area, in P. Andrews and A. Crockett (eds) *Three excavations along the Thames and its tributaries, 1994: Neolithic to Saxon settlement and burial in the Thames, Colne, and Kennet Valleys*: 113-170. Salisbury: Wessex Archaeology.

Crowson, A., T. Lane, K. Penn and D. Trimble 2005. *Anglo-Saxon Settlement on the Siltland of Eastern England*. Sleaford: Heritage Trust of Lincolnshire.

Cunliffe, B.W. 2005. *Iron Age Communities in Britain: an account of England, Scotland and Wales from the seventh century BC until the Roman Conquest*. 4th ed. Abingdon: Routledge.

Dalwood, H., J. Dillon, J. Evans and A. Hawkins 1989. Excavations in Walton, Aylesbury, 1985-1986. *Records of Buckinghamshire* 31: 137–225.

Dark, P. 2000. *The Environment of Britain in the First Millennium AD*. London: Duckworth.

Davies, G. 2008. An archaeological evaluation of the Middle-Late Anglo-Saxon settlement at Chalkpit Field, Sedgeford, Northwest Norfolk. Unpublished draft report, <http://www.scribd.com/doc/3989245/CNEreport-draft>, accessed November 2011.

De'Athe, R. 2012. Early to middle Anglo-Saxon settlement, a lost medieval church rediscovered and an early post-medieval cemetery in Wilton. *Wiltshire Archaeological and Natural History Magazine* 105: 117–144.

Dennell, R. 1972. The interpretation of plant remains: Bulgaria, in E.S. Higgs (ed.) *Papers in Economic Prehistory*: 149-159. Cambridge: Cambridge University Press.

Dennell, R. 1976a. The economic importance of plant resources represented on archaeological sites. *Journal of Archaeological Science* 3: 229–247.

Dennell, R. 1976b. Prehistoric crop cultivation in southern England: a reconsideration. *The Antiquaries Journal* 56: 11–23.

Dodwell, N., S. Lucy and J. Tipper 2004. Anglo-Saxons on the Cambridge Backs: the Criminology site settlement and King's Garden Hostel cemetery. *Proceedings of the Cambridge Antiquarian Society* 93: 95–124.

Edwards, C. 2008. Saxon archaeology and medieval archaeology at Forbury House, Reading. *Berkshire Archaeological Journal* 77: 39–44.

Ellenberg, H. 1988. *Vegetation Ecology of Central Europe*. Cambridge: Cambridge University Press.

Ellis, C.J. 2004. *A Prehistoric Ritual Complex at Eynesbury, Cambridgeshire. Excavation of a Multi-Period Site in the Great Ouse Valley, 2000-2001*. Salisbury: Trust for Wessex Archaeology.

Evans, C., D. Mackay and L. Webley 2008. *Borderlands. The Archaeology of the Addenbrooke's Environs, South Cambridge*. Cambridge: Cambridge Archaeological Unit.

Evenson, R. 1974. International Diffusion of Agrarian Technology. *The Journal of Economic History* 34: 51–73.

Faith, R. 1997. *The English peasantry and the growth of lordship*. London: Leicester University Press.

Faith, R. 2009. Forces and Relations of Production in Early Medieval England. *Journal of Agrarian Change* 9: 23–41.

Farley, M. 1976. Saxon and medieval Walton, Aylesbury: excavations 1973-4. *Records of Buckinghamshire* 20: 153–290.

Fitter, A. and H.J. Peat 1994. The Ecological Flora Database. *Journal of Ecology* 82: 415–425.

Fletcher, T. 2008. Anglo-Saxon Settlement and Medieval Pits at 1 High Street, Willingham, Cambridgeshire. Unpublished report, Cambridgeshire Archaeological Field Unit.

Ford, S. 2002. *Charnham Lane, Hungerford, Berkshire; Archaeological Investigations 1988-1997*. Reading: Thames Valley Archaeological Services.

Ford, S. and I.J. Howell 2004. Saxon and Bronze Age Settlement at the Orchard Site, Walton Road, Walton, Aylesbury, 1994, in S. Ford, I.J. Howell and K. Taylor (eds) *The archaeology of the Aylesbury-Chalgrove gas pipeline and The Orchard, Walton Road, Aylesbury*: 61-88. Reading: Thames Valley Archaeological Services.

Foreman, S., J. Hiller and D. Petts 2002. *Gathering the people, settling the land. The archaeology of a Middle Thames Landscape: Anglo-Saxon to post-medieval*. Oxford: Oxford Archaeology.

Fowler, P.J. 1976. Agriculture and rural settlement, in D.M. Wilson (ed.) *The Archaeology of Anglo-Saxon England*: 23-48. London: Methuen.

Fowler, P.J. 1981. Farming in the Anglo-Saxon landscape: an archaeologist's review. *Anglo-Saxon England* 9: 263-280.

Fowler, P.J. 1999. Agriculture, in M. Lapidge, J. Blair, S. Keynes, and D. Scragg (eds) *The Blackwell Encyclopaedia of Anglo-Saxon England*: 21-23. Oxford: Blackwell Publishing.

Fowler, P.J. 2002. *Farming in the First Millennium AD: British agriculture between Julius Caesar and William the Conqueror*. Cambridge: Cambridge University Press.

Fryer, V. 2008. An assessment of the charred plant macrofossils and other remains from Eye, Suffolk (EYE 083). Unpublished manuscript from J. Caruth, Suffolk County Council Archaeological Service.

Fryer, V. and P. Murphy 1996. Macrobotanical and other remains from Whitehouse Industrial Estate, Ipswich, Suffolk (IPS 247): An Assessment. Unpublished manuscript from J. Caruth, Suffolk County Council Archaeological Service.

Gardiner, M. 2012. Stacks, Barns and Granaries in Early and High Medieval England: Crop Storage and its Implications, in J.A. Quirós Castillo (ed.) *Horrea, Silos and Barns*: 23-38. Vitoria-Gasteiz: University of the Basque Country.

Garrow, D., S. Lucy and D. Gibson 2006. *Excavations at Kilverstone, Norfolk: an Episodic Landscape History*. Cambridge: Cambridge Archaeological Unit.

Gibson, C. 2003. An Anglo-Saxon Settlement at Godmanchester, Cambridgeshire. *Anglo-Saxon Studies in Archaeology and History* 12: 137–217.

Goldberg, P. and R. Macphail 2006. *Practical and theoretical geoarchaeology*. Oxford: Blackwell Publishing.

Green, F. 1981. Iron Age, Roman and Saxon Crops: The Archaeological Evidence from Wessex, in M. Jones and G. Dimbleby (eds) *The Environment of Man: the Iron Age to the Anglo-Saxon Period*: 129-153. Oxford: British Archaeological Reports.

Green, F. 1982. Problems of interpreting differentially preserved plant remains from excavations of medieval urban sites, in A. Hall and H. Kenward (eds) *Environmental Archaeology in the Urban Context*: 40-46. London: Council for British Archaeology.

Grime, J.P., J.G. Hodgson and R. Hunt 1988. *Comparative Plant Ecology: a functional approach to common British species*. London: Unwin Hyman.

Hagen, A. 2006. *Anglo-Saxon Food and Drink. Production, Processing, Distribution and Consumption*. Hockwold cum Wilton: Anglo-Saxon Books.

Hall, R.V. 2003. Archaeological evaluation of land north of the Post Office, Church Terrace, Outwell, Norfolk. Unpublished report, Archaeological Project Services.

Hamerow, H. 1999. Settlement patterns, in M. Lapidge, J. Blair, S. Keynes and D. Scragg (eds) *The Blackwell Encyclopaedia of Anglo-Saxon England*: 416-418. Oxford: Blackwell Publishing.

Hamerow, H. 2002. *Early medieval settlements: the archaeology of rural communities in Northwest Europe, 400-900*. Oxford: Oxford University Press.

Hamerow, H. 2007. Agrarian production and the emporia of mid Saxon England, ca.AD 650-850, in J. Henning (ed.) *Post-Roman Towns, Trade and Settlement in Europe and Byzantium. Vol. 1. The Heirs of the Roman West*: 219-232. Berlin: Walter de Gruyter.

Hamerow, H. 2011. Overview: Rural Settlement, in H. Hamerow, D.A. Hinton and S. Crawford (eds) *The Oxford Handbook of Anglo-Saxon Archaeology*: 119-127. Oxford: Oxford University Press.

Hamerow, H. 2012. *Rural Settlements and Society in Anglo-Saxon England*. Oxford: Oxford University Press.

Hamerow, H. 2017. Feeding Anglo-Saxon England: the bioarchaeology of an agricultural revolution ('FeedSax'). *Medieval Settlement Research* 32: 85–86.

Hamerow, H., C. Hayden and G. Hey 2007. Anglo-Saxon and Earlier Settlement near Drayton Road, Sutton Courtenay, Berkshire. *The Archaeological Journal* 164: 109–196.

Hancock, A. 2010. Excavation of a Mid-Saxon Settlement at Water Eaton, Bletchley, Milton Keynes. *Records of Buckinghamshire* 50: 5–24.

Hardy, A., A. Dodd and G.D. Keevill 2003. *Ælfric's Abbey: Excavations at Eynsham Abbey*. Oxford: Oxford Archaeology.

Havis, R. and H. Brooks 2004. *Excavations at Stansted Airport, 1986-91*. 2 volumes. Chelmsford: Essex County Council.

Henning, J. 2014. Did the "agricultural revolution" go east with Carolingian conquest? Some reflections on early medieval rural economics of the Baiuvarii and Thuringi, in J. Fries-Knoblach, H. Steuer and J. Hines (eds) *Baiuvarii and Thuringi: An Ethnographic Perspective*: 331-359. Woodbridge: Boydell Press.

Hey, G. 2004. *Yarnton: Saxon and Medieval Settlement and Landscape. Results of Excavations 1990-96*. Oxford: Oxford Archaeology.

Hill, D. 2000. Sulh - the Anglo-Saxon plough c.1000 AD. *Landscape History* 22: 7–19.

Hill, M.O., C.D. Preston and D.B. Roy 2004. *PLANTATT. Attributes of British and Irish Plants: Status, Size, Life History, Geography and Habitats*. Huntingdon: Centre for Ecology and Hydrology.

Hillman, G. 1981. Reconstructing Crop Husbandry Practices from Charred Remains of Crops, in R. Mercer (ed.) *Farming Practice in British Prehistory*: 123-162. Edinburgh: Edinburgh University Press.

Hillman, G. 1984. Interpretation of archaeological plant remains: the application of ethnographic models from Turkey, in W. van Zeist and W.A. Casparie (eds) *Plants and ancient man*: 1-42. Rotterdam: Balkema.

Hillman, G., S. Mason, D. de Moulins and M. Nesbitt 1996. Identification of archaeological remains of wheat: the 1992 London workshop. *Circaea* 12: 195–209.

Hodges, R. 1989. *The Anglo-Saxon Achievement: Archaeology and the beginnings of English society*. London: Duckworth.

Holmes, M. 2014. *Animals in Saxon and Scandinavian England: Backbones of Economy and Society*. Leiden: Sidestone Press.

Hood, A. 2007. Alpha Park, Great North Road, East Socon, Cambridgeshire: Archaeological Strip, Map and Sample: Post Excavation Assessment. Unpublished report, Foundations Archaeology.

Hubbard, R.N.L. 1980. Development of Agriculture in Europe and the near East: Evidence from Quantitative Studies. *Economic Botany* 34: 51–67.

Hughes, M.K. and H.F. Diaz 1994. Was there a "Medieval Warm Period", and if so, where and when? *Climatic Change* 26: 109–142.

Jackson, R.P.J. and T.W. Potter 1996. *Excavations at Stonea, Cambridgeshire 1980-85*. London: British Museum.

Jacomet, S. 2006. *Identification of cereal remains from archaeological sites*. 2nd ed. Basel: Basel University.

Jessen, K. and H. Helbaek 1944. *Cereals in Great Britain and Ireland in Prehistoric and Early Historic Times*. Copenhagen: Munksgaard.

Jones, A.K.G. Wicken Bonhunt Plant Remains. A preliminary report. Ancient Monuments Laboratory Report 1760.

Jones, G. 1984. Interpretation of archaeological plant remains: ethnographic models from Greece, in W. van Zeist and W.A. Casparie (eds) *Plants and Ancient Man*: 43-61. Rotterdam: Balkema.

Jones, G. 1987. A statistical approach to the identification of crop processing. *Journal of Archaeological Science* 14: 311–323.

Jones, G. 1990. The application of present-day cereal processing studies to charred archaeobotanical remains. *Circaea* 6: 91–96.

Jones, G. 1991. Numerical analysis in archaeobotany, in W. van Zeist, K. Wasylikowa and K.E. Behre (eds) *Progress in Old World Palaeoethnobotany*: 63-80. Rotterdam: Balkema.

Jones, G. 1992. Weed phytosociology and crop husbandry: identifying a contrast between ancient and modern practice. *Review of Palaeobotany and Palynology* 73: 133–143.

Jones, G., M. Charles, A. Bogaard and J.G. Hodgson 2010. Crops and weeds: the role of weed functional ecology in the identification of crop husbandry methods. *Journal of Archaeological Science* 37: 70–77.

Jones, M. 1981. The Development of Crop Husbandry, in M. Jones and G. Dimbleby (eds) *The Environment of Man: the Iron Age to the Anglo-Saxon Period*: 95-127. Oxford: British Archaeological Reports.

Jones, M. 1988. The Arable Field: A Botanical Battleground, in M. Jones (ed.) *Archaeology and the Flora of the British Isles. Human influence on the evolution of plant communities*: 86-92. Oxford: Oxford University Committee for Archaeology.

Jones, M. 2009. Dormancy and the plough: Weed seed biology as an indicator of agraian change in the first millennium AD, in A. Fairbairn and E. Weiss (eds) *From Foragers to Farmers. Papers in honour of Gordon C. Hillman*: 58-63. Oxford and Oakville: Oxbow.

Kay, Q.O.N. 1971. Anthemis Cotula L. *Journal of Ecology* 59: 623–636.

Kenney, S. 2002. Roman, Saxon and Medieval Occupation at the site of the former Red, White and Blue Public House, Chiefs Street, Ely. Unpublished report, Cambridgeshire County Council.

Kenyon, D. and M. Watts 2006. An Anglo-Saxon enclosure at Copsehill Road, Lower Slaughter: excavations in 1999. *Transactions of the Bristol and Gloucestershire Archaeological Society* 124: 73–109.

Lambrick, G. 2010. *Neolithic to Saxon social and environmental change at Mount Farm, Berinsfield, Dorchester-on-Thames*. Oxford: Oxford Archaeology.

Launert, E. 1981. *The Hamlyn Guide to Edible and Medicinal Plants of Britain and Northern Europe*. London: Hamlyn Publishing Group.

Lavender, N.J. 1998. A Saxon building at Chadwell St. Mary: excavations at Chadwell St. Mary County Primary School 1996. *Essex Archaeology and History* 29: 48-58.

Lawson, A.J. 1983. *The Archaeology of Witton, near North Walsham, Norfolk*. Dereham: Norfolk Museums Service.

Lebecq, S. 2000. The role of monasteries in the systems of production and exchange of the Frankish world between the seventh and the beginning of the ninth centuries, in I. Hansen and C. Wickham (eds) *The Long Eighth Century*: 121-148. Leiden: Brill.

Lovell, J., J. Timby, G. Wakeham and M.J. Allen 2007. Iron-Age to Saxon Farming Settlement at Bishop's Cleeve, Gloucestershire: excavations south of Church Road, 1998 and 2004. *Transactions of the Bristol and Gloucestershire Archaeological Society* 125: 95-129.

Lucy, S., J. Tipper and A. Dickens 2009. *The Anglo-Saxon Settlement and Cemetery at Bloodmoor Hill, Carlton Colville, Suffolk*. Cambridge: Cambridge Archaeological Unit.

Malcolm, G., D. Bowsher and R. Cowie 2003. *Middle Saxon London. Excavations at the Royal Opera House 1989-99*. London: Museum of London Archaeology.

Masefield, R. 2008. *Prehistoric and later settlement and landscape from Chiltern Scarp to Aylesbury Vale: the archaeology of the Aston Clinton Bypass, Buckinghamshire*. Oxford: British Archaeological Reports.

McKerracher, M. 2014a. Agricultural Development in Mid Saxon England. Unpublished DPhil dissertation, University of Oxford.

McKerracher, M. 2014b. Landscapes of Production in Mid Saxon England: the monumental grain ovens. *Medieval Settlement Research* 29: 82-85.

McKerracher, M. 2015. Assessment of macroscopic plant remains from the 2014 excavations at Lyminge, Kent. Unpublished report for University of Reading.

McKerracher, M. 2016. Bread and surpluses: the Anglo-Saxon "bread wheat thesis" reconsidered. *Environmental Archaeology* 21: 88-102.

McKerracher, M. 2017. Seeds and status: the archaeobotany of monastic Lyminge, in G. Thomas and A. Knox (eds) *Early medieval monasticism in the North Sea Zone*: 127-134. Oxford: University of Oxford School of Archaeology.

McKerracher, M. 2018. *Farming Transformed in Anglo-Saxon England: Agriculture in the Long Eighth Century*. Oxford: Windgather.

Miles, D. 1986. *Archaeology at Barton Court Farm, Abingdon, Oxon. An investigation of late Neolithic, Iron Age, Romano-British, and Saxon settlements*. Oxford and London: Oxford Archaeological Unit and the Council for British Archaeology.

Moffett, L. 1991. The archaeobotanical evidence for free-threshing tetraploid wheat in Britain, in E. Hajnalová (ed.) *Palaeoethnobotany and archaeology. International Work-Group for Palaeoethnobotany 8th Symposium, Nitra-Nové Vozokany 1989*: 233-243. Nitra: Archaeological Institute of the Slovak Academy of Sciences.

Moffett, L. 2006. The Archaeology of Medieval Plant Foods, in C.M. Woolgar, D. Serjeantson and T. Waldron (eds) *Food in Medieval England*: 41-55. Oxford: Oxford University Press.

Moffett, L. 2011. Food plants on archaeological sites: the nature of the archaeobotanical record, in H. Hamerow , D.A. Hinton and S. Crawford (eds) *The Oxford Handbook of Anglo-Saxon Archaeology*: 346-360. Oxford: Oxford University Press.

Monk, M. 1977. The Plant Economy and Agriculture of the Anglo-Saxons in Southern Britain: with particular reference to the "mart" settlements at Southampton and Winchester. Unpublished Masters dissertation, University of Southampton.

Moore, J. 2001. Excavations at Oxford Science Park, Littlemore, Oxford. *Oxoniensia* 66: 163–219.

Morrison, K.D. 1994. The Intensification of Production: Archaeological Approaches. *Journal of Archaeological Method and Theory* 1: 111–159.

Mortimer, R. 1996. Excavations of a group of Anglo-Saxon features at Denny End, Waterbeach, Cambridgeshire. Unpublished report, Cambridge Archaeological Unit.

Mortimer, R. 1998. Excavation of the Middle Saxon to Medieval village at Lordship Lane, Cottenham, Cambridgeshire. Unpublished report, Cambridge Archaeological Unit.

Mortimer, R. 2003. Rosemary Lane, Church End, Cherry Hinton. Unpublished report, Cambridge Archaeological Unit.

Mortimer, R. and C. Evans 1996. An Archaeological Excavation at Hinxton Quarry, Cambridgeshire, 1995. 2 volumes. Unpublished report, Cambridge Archaeological Unit.

Mortimer, R., R. Regan and S. Lucy 2005. *The Saxon and Medieval Settlement at West Fen Road, Ely: The Ashwell Site*. Cambridge: Cambridge Archaeological Unit.

Mudd, A. 2002. *Excavations at Melford Meadows, Brettenham, 1994: Romano-British and Early Saxon Occupations*. Oxford: Oxford Archaeological Unit.

Mudd, A. 2007. *Bronze Age, Roman and Later Occupation at Chieveley, West Berkshire. The archaeology of the A34/M4 Road Junction Improvement.* Oxford: British Archaeological Reports.

Mudd, A. and M. Webster 2011. *Iron Age and Middle Saxon Settlements at West Fen Road, Ely, Cambridgeshire: The Consortium Site.* Oxford: British Archaeological Reports.

Murphy, P. 1994. The Anglo-Saxon landscape and rural economy: some results from sites in East Anglia and Essex, in J. Rackham (ed.) *Environment and Economy in Anglo-Saxon England*: 23-39. York: Council for British Archaeology.

Murphy, P. 2004. The Environment and Agarian Economy of Saxon and Medieval Ipswich. Unpublished manuscript from author.

Murphy, P. 2010. The Landscape and Economy of the Anglo-Saxon Coast: New Archaeological Evidence, in N.J. Higham and M. J. Ryan (eds) *Landscape Archaeology of Anglo-Saxon England*: 211-221. Woodbridge: Boydell Press.

Murray, J. 2005. Excavations at Station Road, Gamlingay, Cambridgeshire. *Anglo-Saxon Studies in Archaeology and History* 13: 173–330.

Mynard, D.C. and R.J. Zeepvat 1992. *Excavations at Great Linford, 1974-80.* Aylesbury: Buckinghamshire Archaeological Society.

Newman, J. 1992. The late Roman and Anglo-Saxon settlement pattern in the Sandlings of Suffolk, in M. Carver (ed.) *The Age of Sutton Hoo: The Seventh Century in North-Western Europe*: 25-38. Woodbridge: Boydell Press.

Newton, A.A.S. 2010. Saxon and medieval settlement at The Old Bell, Marham, Norfolk. Research Archive Report. Unpublished report, Archaeological Solutions.

Norton, G.W. and J. Alwang 1993. *Introduction to Economics of Agricultural Development.* New York: McGraw-Hill.

O'Brien, L. 2016. *Bronze Age Barrow, Early to Middle Iron Age Settlement and Burials, and Early Anglo-Saxon Settlement at Harston Mill, Cambridgeshire.* Bury St Edmunds: Archaeological Solutions.

Oosthuizen, S. 2005. New Light on the Origins of Open-field Farming? *Medieval Archaeology* 49: 165–193.

Oosthuizen, S. 2013. *Tradition and Transformation in Anglo-Saxon England: Archaeology, Common Rights and Landscape.* London: Bloomsbury Academic.

Patrick, C. and S. Rátkai 2011. Hillside Meadow, Fordham, in R. Cuttler, H. Martin-Bacon, K. Nichol, C. Patrick, R. Perrin, S. Rátkai, M. Smith and J. Williams (eds) *Five Sites in Cambridgeshire. Excavations at Woodhurst, Fordham, Soham, Buckden and St. Neots, 1998-2002*: 41-123. Oxford: British Archaeological Reports.

Pelling, R. 2003. Early Saxon cultivation of emmer wheat in the Thames Valley and its cultural implications, in K.A. Robson Brown (ed.) *Archaeological Sciences 1999. Proceedings of the Archaeological Sciences Conference, University of Bristol, 1999*: 103-110. Oxford: British Archaeological Reports.

Pelling, R. and M. Robinson 2000. Saxon Emmer Wheat from the Upper and Middle Thames Valley, England. *Environmental Archaeology* 5: 117–119.

Phillips, M. 2005. Excavations of an Early Saxon settlement at Pitstone. *Records of Buckinghamshire* 45: 1–32.

Pine, J. 2001. The excavation of a Saxon settlement at Cadley Road, Collingbourne Ducis, Wiltshire. *Wiltshire Archaeological and Natural History Magazine* 94: 88–117.

Pine, J. 2009. Latton Quarry, Latton, Wiltshire. A Post-Excavation Assessment for Hills Quarry Products. Unpublished report, Thames Valley Archaeological Services.

Pine, J. and S. Ford 2003. Excavation of Neolithic, Late Bronze Age, Early Iron Age and Early Saxon Features at St. Helen's Avenue, Benson, Oxfordshire. *Oxoniensia* 68: 131–178.

Pollard, A.M. 2009. Measuring the passage of time: achievements and challenges in archaeological dating, in B. Cunliffe, C. Gosden and R.A. Joyce (eds) *The Oxford Handbook of Archaeology*: 145-168. Oxford: Oxford University Press.

Pollard, J. 1996. Excavations at Bourn Bridge, Pampisford, Cambridgeshire: Part 2, Roman and Saxon. Unpublished report, Cambridge Archaeological Unit.

Preston, S. 2007. Bronze Age Occupation and Saxon Features at the Wolverton Turn Enclosure, near Stony Stratford, Milton Keynes: investigations by Tim Schadla-Hall, Philip Carstairs, Jo Lawson, Hugh Beamish, Andrew Hunn, Ben Ford and Tess Durden, 1972 to 1994. *Records of Buckinghamshire* 47: 81–117.

Puy, A. and A.L. Balbo 2013. The genesis of irrigated terraces in al-Andalus. A geoarchaeological perspective on intensive agriculture in semi-arid environments (Ricote, Murcia, Spain). *Journal of Arid Environments* 89: 45–56.

Reynolds, A. 2003. Boundaries and settlements in later sixth to eleventh century England, in D. Griffiths, A. Reynolds and S. Semple (eds) *Boundaries in Early Medieval Britain* (Anglo-Saxon Studies in Archaeology and History 12): 98-136. Oxford: Oxford University School of Archaeology.

Rickett, R. 1995. *The Anglo-Saxon Cemetery at Spong Hill, North Elmham, Part VII: The Iron Age, Roman and Early Saxon Settlement*. Gressenhall: Norfolk Museums Service.

Rippon, S. 2010. Landscape Change during the "Long Eighth Century" in Southern England, in N.J. Higham and M.J. Ryan (eds) *The Landscape Archaeology of Anglo-Saxon England*: 39-64. Woodbridge: Boydell Press.

Rippon, S., R.M. Fyfe and A.G. Brown 2006. Beyond Villages and Open Fields: The Origins and Development of a Historic Landscape Characterised by Dispersed Settlement in South-West England. *Medieval Archaeology* 50: 31–70.

Rippon, S., C. Smart and B. Pears 2015. *The fields of Britannia: continuity and change in the late Roman and early medieval landscape*. Oxford: Oxford University Press.

Robinson, M. Assessment of Carbonised Plant Remains from Rivers Nightclub, Benson, Oxon (SAB99114). Unpublished report on file at Oxfordshire Historic Environment Records.

Robinson, M. 1992. Environment, archaeology and alluvium on the river gravels of the South Midlands, in S. Needham and M.G. Macklin (eds) *Alluvial Archaeology in Britain*: 197-208. Oxford: Oxbow.

Robinson, M. 1997. Charred plant remains from Walton, Aylesbury, Bucks. Unpublished report on file at Buckinghamshire Historic Environment Records.

Rowley, T. and M. Steiner 1996. *Cogges Manor Farm Witney Oxfordshire. The excavations from 1986-1994 and the Historic Building Analysis*. Oxford: University of Oxford and Oxfordshire County Council.

Ruas, M.-P. 2005. Aspects of early medieval farming from sites in Mediterranean France. *Vegetation History and Archaeobotany* 14: 400–415.

Ruttan, V.W. 1998. Models of Agricultural Development, in C.K. Eicher and J.M. Staatz (eds) *International agricultural development*: 155-162. Baltimore: Johns Hopkins University Press.

Spedding, C.R.W., J.M. Walsingham and A.M. Hoxey 1981. *Biological efficiency in agriculture*. London and New York: Academic Press.

Stace, C.A. 2010. *New Flora of the British Isles*. 3rd ed. Cambridge: Cambridge University Press.

Stokes, P. and P. Rowley-Conwy 2002. Iron Age Cultigen? Experimental Return Rates for Fat Hen (Chenopodium album L.). *Environmental Archaeology* 7: 95–99.

Stone, P. 2011. Saxon and Medieval Activity at Walton Street, Aylesbury. *Records of Buckinghamshire* 51: 99–129.

Streeter, D. 2010. *Collins Flower Guide*. London: Collins.

Suffolk County Council Archaeological Service 2015. *Ipswich 1974-1990 Excavation Archive*. Online resource accessed February 2019, published by the Archaeology Data Service, <https://doi.org/10.5284/1034376>.

Taylor, K. and S. Ford 2004. Late Bronze Age, Iron Age, Roman and Saxon sites along the Oxfordshire section, in S. Ford, I.J. Howell and K. Taylor (eds) *The archaeology of the Aylesbury-Chalgrove gas pipeline and The Orchard, Walton Road, Aylesbury*: 25-58. Reading: Thames Valley Archaeological Services.

Tester, A., S. Anderson, I. Riddler and R. Carr 2014. *Staunch Meadow, Brandon, Suffolk: a high status Middle Saxon settlement on the fen edge*. Bury St Edmunds: Suffolk County Council Archaeological Service.

Thomas, G., G. McDonnell, J. Merkel and P. Marshall 2016. Technology, ritual and Anglo-Saxon agrarian production: the biography of a seventh-century plough coulter from Lyminge, Kent. *Antiquity* 90: 742–758.

Tipper, J. 2004. *The Grubenhaus in Anglo-Saxon England: an analysis and interpretation of the evidence from a most distinctive building type*. Yedingham: Landscape Research Centre.

Tipper, J. 2007. West Stow, Lackford Bridge Quarry (WSW 030). A report on a rescue excavation undertaken in 1978-9. Unpublished report, Suffolk County Council Archaeological Service.

Ulmschneider, K. 2000. *Markets, Minsters, and Metal-Detectors. The archaeology of Middle Saxon Lincolnshire and Hampshire compared*. Oxford: British Archaeological Reports.

Ulmschneider, K. 2011. Settlement Hierarchy, in H. Hamerow, D.A. Hinton and S. Crawford (eds) *The Oxford Handbook of Anglo-Saxon Archaeology*: 156-171. Oxford: Oxford University Press.

Vaughan-Williams, A. 2005. Report on the plant remains from Forbury House Reading, in C. Edwards and S. Adams (eds) Forbury House, Reading, Berkshire. Archive Report: 44-48. Unpublished report, AOC Archaeology Group.

van der Veen, M. 1989. Charred Grain Assemblages from Roman-Period Corn Driers in Britain. *The Archaeological Journal* 146: 302–319.

van der Veen, M. 1992. *Crop Husbandry Regimes: an Archaeobotanical Study of Farming in Northern England, 1000 BC - AD 500*. Sheffield: J.R. Collis.

van der Veen, M. 2007. Formation processes of desiccated and carbonized plant remains - the identification of routine practice. *Journal of Archaeological Science* 34: 968–990.

van der Veen, M. 2010. Agricultural innovation: invention and adoption or change and adaptation? *World Archaeology* 42: 1–12.

van der Veen, M., A. Hill and A. Livarda 2013. The Archaeobotany of Medieval Britain (c AD 450-1500): Identifying Research Priorities for the 21st Century. *Medieval Archaeology* 57: 151–182.

van der Veen, M. and G. Jones 2006. A re-analysis of agricultural production and consumption: implications for understanding the British Iron Age. *Vegetation History and Archaeobotany* 15: 217–228.

Wade, K. 1980. A settlement site at Bonhunt Farm, Wicken Bonhunt, Essex, in D.G. Buckley (ed.) *Archaeology in Essex to AD 1500*: 96-102. London: Council for British Archaeology.

Wall, W. 2011. Middle Saxon Iron Smelting near Bonemills Farm, Wittering, Cambridgeshire. *Anglo-Saxon Studies in Archaeology and History* 17: 87–100.

Wallis, S. and M. Waughman 1998. *Archaeology and the Landscape in the Lower Blackwater Valley*. Chelmsford: Essex County Council.

Watts, M. 2002. *The archaeology of mills and milling*. Stroud: Tempus Publishing.

Webster, C.J. 2008. *The Archaeology of South West England*. Taunton: Somerset Heritage Service.

Welch, M. 2011. The Mid Saxon "Final Phase", in H. Hamerow, D.A. Hinton and S. Crawford (eds) *The Oxford Handbook of Anglo-Saxon Archaeology*: 266-287. Oxford: Oxford University Press.

West, S.E. 1985. *West Stow. The Anglo-Saxon Village.* 2 volumes. Ipswich: Suffolk County Council.

Whitelock, D. 1979. *English Historical Documents, Vol. I: 500-1042*. 2nd ed. London: Methuen.

Williams, G. 2008. An Archaeological Evaluation at Milton Park, Didcot, Oxfordshire. Unpublished report, John Moore Heritage Servies.

Williams, P. and R. Newman 2006. *Market Lavington, Wiltshire, An Anglo-Saxon Cemetery and Settlement. Excavations at Grove Farm, 1986-90*. Salisbury: Wessex Archaeology.

Williams, R.J. 1993. *Pennyland and Hartigans. Two Iron Age and Saxon sites in Milton Keynes*. Aylesbury: Buckinghamshire Archaeological Society.

Williams, R.J., P.J. Hart and A.T.L. Williams 1996. *Wavendon Gate. A Late Iron Age and Roman settlement in Milton Keynes.* Aylesbury: Buckinghamshire Archaeological Society.

Williams, R.J. and R.J. Zeepvat 1994. *Bancroft. A Late Bronze Age/Iron Age Settlement, Roman Villa and Temple-Mausoleum.* 2 volumes. Aylesbury: Buckinghamshire Archaeological Society.

Williamson, T. 2003. *Shaping medieval landscapes.* Macclesfield: Windgather.

Wright, J. 2004. Anglo-Saxon Settlement at Cherry Orton Road, Orton Waterville, Peterborough. Report on the 2003 archaeological excavations. Unpublished report, Wessex Archaeology.

Wright, J., M. Leivers, R. Seager Smith and C.J. Stevens 2009. *Cambourne New Settlement. Iron Age and Romano-British settlement on the clay uplands of west Cambridgeshire.* Salisbury: Wessex Archaeology.

Zohary, D., M. Hopf and E. Weiss 2012. *Domestication of Plants in the Old World. The origin and spread of domesticated plants in south-west Asia, Europe, and the Mediterranean Basin.* 4th ed. Oxford: Oxford University Press.